Lecture Notes in Computer Science 11310

Commenced Publication in 1973
Founding and Former Series Editors:
Gerhard Goos, Juris Hartmanis, and Jan van Leeuwen

More information about this series at http://www.springer.com/series/8637

Abdelkader Hameurlain · Roland Wagner
Djamal Benslimane · Ernesto Damiani
William I. Grosky (Eds.)

Transactions on Large-Scale Data- and Knowledge-Centered Systems XXXIX

Special Issue on Database- and Expert-Systems Applications

 Springer

Editors-in-Chief

Abdelkader Hameurlain
IRIT, Paul Sabatier University
Toulouse, France

Roland Wagner
FAW, University of Linz
Linz, Austria

Guest Editors

Djamal Benslimane ⓘ
IUT, University Lyon 1
Villeurbanne Cedex, France

William I. Grosky
University of Michigan-Dearborn
Dearborn, MI, USA

Ernesto Damiani ⓘ
University of Milan
Crema, Italy

ISSN 0302-9743 ISSN 1611-3349 (electronic)
Lecture Notes in Computer Science
ISSN 1869-1994 ISSN 2510-4942 (electronic)
Transactions on Large-Scale Data- and Knowledge-Centered Systems
ISBN 978-3-662-58414-9 ISBN 978-3-662-58415-6 (eBook)
https://doi.org/10.1007/978-3-662-58415-6

Library of Congress Control Number: 2018961592

This Springer imprint is published by the registered company Springer-Verlag GmbH, DE part of Springer Nature
The registered company address is: Heidelberger Platz 3, 14197 Berlin, Germany

Preface

The 28th International Conference on Database and Expert Systems Applications—DEXA 2017—took place in Lyon, France, during August 28–31, 2017. DEXA is an annual conference, designed for researchers interested in some aspects of database, information, and knowledge systems. DEXA 2017 attracted 166 submissions from all over the world.The DEXA 2017 chairs accepted 37 full research papers and 37 short research papers yielding an acceptance rate of 22% for each type of accepted papers.

Among the 37 full papers, we selected 11 papers for their quality and asked authors to extend their papers for a possible inclusion in the TLDKS journal. Finally, only seven papers were accepted and are included in this volume.

The success of this TLDKS special issue of DEXA 2017 would not have been possible without the help of Gabriela Wagner as manager of the DEXA organization, and the professional and hard work of the reviewers. We gratefully thank them for their commitment to this scientific event.

For the readers of this volume, we hope you will find it interesting. We also hope it will inspire you to greater achievements.

September 2018

Djamal Benslimane
Ernesto Damiani
William I. Grosky

Organization

Editorial Board

Contents

Querying Interlinked Data by Bridging RDF Molecule Templates

Kemele M. Endris[1]([✉]), Mikhail Galkin[2,5], Ioanna Lytra[2,4],
Mohamed Nadjib Mami[2], Maria-Esther Vidal[1,2,3], and Sören Auer[1,2,3]

[1] L3S Research Center, University of Hannover, Hannover, Germany
{endris,vidal,auer}@L3S.de
[2] Fraunhofer IAIS, Sankt Augustin, Germany
{galkin,lytra,mami}@iais.fraunhofer.de
[3] Leibniz Information Centre For Science and Technology University Library (TIB),
Hannover, Germany
[4] University of Bonn, Bonn, Germany
[5] ITMO University, Saint Petersburg, Russia

Abstract. Linked Data initiatives have encouraged the publication of a
large number of RDF datasets created by different data providers independently. These datasets can be accessed using different Web interfaces, e.g., SPARQL endpoint; however, federated query engines are still
required in order to provide an integrated view of these datasets. Given
the large number of Web accessible RDF datasets, SPARQL federated
query engines implement query processing techniques to effectively select
the relevant datasets that provide the data required to answer a query.
Existing federated query engines usually utilize coarse-grained description methods where datasets are characterized based on their vocabularies or schema, and details about data in the dataset are ignored,
e.g., classes, properties, or relations. This lack of source description may
lead to the erroneous selection of data sources for a query, and unnecessary retrieval of data and source communication, affecting thus the
performance of query processing over the federation. We address the
problem of federated SPARQL query processing and devise MULDER,
a query engine for federations of RDF data sources. MULDER describes
data sources in terms of an abstract description of entities belonging
to the same RDF class, dubbed as an *RDF molecule template*, and utilizes them for source selection, and query decomposition and optimization. We empirically study the performance and continuous efficiency of
MULDER on existing benchmarks, and compare with respect to existing
federated SPARQL query engines. The experimental results suggest that
RDF *molecule templates* empower MULDER, and allow for selection of
RDF data sources that not only reduce execution time, but also increase
answer completeness and continuous efficiency of MULDER.

1 Introduction

A wide range of communities, e.g., scientific, governmental, and industrial,
encouraged by the Linked Data initiatives, have made publicly available their

© Springer-Verlag GmbH Germany, part of Springer Nature 2018
A. Hameurlain et al. (Eds.): TLDKS XXXIX, LNCS 11310, pp. 1–42, 2018.
https://doi.org/10.1007/978-3-662-58415-6_1

data as RDF datasets; as a result, the number of Web accessible RDF datasets has increased considerably in the last ten years [13]. Web access interfaces, e.g., SPARQL endpoints [11] or Linked Data Fragment (LDF) clients [41] enable the access of these RDF data sources. Albeit individually valuable, the integration of these RDF datasets –logical or physical– empowers their value, and allows for uncovering unknown patterns and collecting relevant insights [23].

Federated query engines provide a flexible solution to the problem of query processing over a federation of datasets logically integrated. This problem has been extensively studied by the database [1,8,9,21,44] and semantic web research [2,7,15,25,34,40] communities. Since the number of potentially relevant data sources for a query can be very large, one of the major challenges of these federated query engines is the selection of the minimal number of sources that can provide the data required to completely answer the query. Current approach resort to source descriptions for identifying the relevant sources of a query. However, the majority of existing approaches only collect coarse-grained source descriptions, e.g., vocabularies or schema utilized in the dataset for modeling the data, and ignore fine-grained characteristics, e.g., classes, properties, and relations. Nevertheless, we deem that fine-grained source descriptions represent building blocks not only for effectively selecting relevant sources, but also for identifying query execution plans that will collect the query answers efficiently. Hence, we devise MULDER an integrated approach for federated query processing, which utilizes *RDF Molecule Templates* (RDF-MTs) for a fine-grained description of an RDF dataset, as well as its connections with other datasets.

MULDER exploits RDF-MTs during source selection, and query decomposition and optimization, and is able to produce query plans that answer federated queries effectively. Furthermore, MULDER query plans benefit the continuous generation of query answers. This paper extends our previous work [10], where the concept of *RDF Molecule Templates* is defined; we present a more detailed definition and comparison of the techniques implemented in MULDER. Additionally, it is reported a more comprehensive empirical evaluation of the performance of MULDER in three different benchmarks: the Berlin SPARQL Benchmark (BSBM) [6], FedBench [36]; and LSLOD [19].

Our contributions can be summarized as follows: (1) A thorough formalization accompanied by an implementation for federated query processing employing RDF Molecule Templates (RDF-MTs) for selecting relevant sources, query decomposition, and execution. (2) An analysis of dataset characteristics in a federation using RDF-MTs. (3) An empirical evaluation assessing the performance of MULDER in terms of query execution time and query answer completeness. The reported results provide evidence that MULDER is able to speed up query execution and enhance answer completeness with respect to the state-of-the-art. (4) An experimental study of the continuous efficiency of MULDER in terms of novel metrics are reported, e.g., $dief@t$ and $dief@k$ [3]; observed results suggest that MULDER performance increases gradually and is competitive to the state-of-the-art adaptive federated query engine ANAPSID [2].

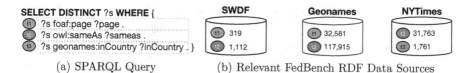

(a) SPARQL Query   (b) Relevant FedBench RDF Data Sources

Fig. 1. Motivating example. (a) SPARQL query over FedBench RDF data sources. (b) FedBench data sources able to execute the query triple patterns. Each triple pattern can be executed in more than one RDF data source.

The remainder of the paper is structured as follows: We describe the behavior of existing systems with a motivating example in Subsect. 1.1 and discuss the preliminaries in Subsect. 1.2. We present the related works and compare them to our approach in Sect. 2. We formally define the query decomposition problem and introduce the MULDER approach in Sect. 3. In Sect. 4, we present the MULDER framework where we get into the details of the Web interface description model of RDF-MTs (Subsect. 4.1), query decomposition techniques (Subsect. 4.2), and query optimization strategies (Subsect. 4.3) used in MULDER. We present the results of our experimental study in Sect. 5 and conclude with an outlook on future work in Sect. 6.

1.1 Motivating Example

We motivate our work by comparing the performance of state-of-the-art federated SPARQL query engines on a federation of RDF data sources from the *FedBench* benchmark. FedBench [37] is a standard benchmark for evaluating federated query processing approaches. It comprises three RDF data collections of interlinked datasets, i.e., the *Cross Domain Collection* (LinkedMDB, DBpedia, GeoNames, NYTimes, SWDF, and Jamendo), the *Life Science Collection* (Drugbank, DBpedia, KEGG, and ChEBI), and the synthetic SP^2Bench *Data Collection*. Although these datasets are from different domains, some RDF vocabularies are utilized in more than one dataset. For instance, foaf properties are used in DBpedia, GeoNames, SWDF, LinkedMDB, and NYTimes, while RDF triples with the owl:sameAs property are present in all the FedBench datasets.

Federated SPARQL query engines, e.g., ANAPSID [2] and FedX [40], provide a unified view of the federation of FedBench datasets, and support query processing over this unified view. Figure 1a presents a SPARQL query on a FedBench federation of three data sources. The query comprises three triple patterns: t1 can be answered on SWDF and Geonames; NYTimes can answer t3, while t2 can be executed over SWDF, Geonames, and NYTimes respectively. Figure 1b reports on the number of answers of t1, t2, and t3 over SWDF, Geonames, and NYTimes. Federated query engines rely on source descriptions to select relevant sources for a query. For instance, based on the vocabulary properties utilized in each of the data sources, ANAPSID decides that SWDF, Geonames, and NYTimes are the relevant sources, while FedX contacts each of the federation SPARQL endpoints to determine where t1, t2, and t3 will be executed.

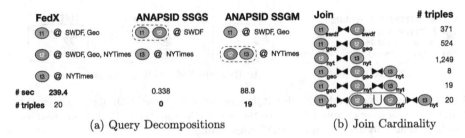

(a) Query Decompositions   (b) Join Cardinality

Fig. 2. Motivating example. (a) Query decompositions by FedX and ANAPSID. (b) Cardinality of joins of triple patterns over relevant RDF data sources. FedX decomposition produces complete answers, but at the cost of execution time. ANAPSID decompositions run faster, but produce incomplete results.

Furthermore, different criteria are followed to decompose the query into the subqueries that will be posed over the relevant sources to collect the data required to answer the query. As presented in Fig. 2a, FedX identifies that t3 composes an exclusive group and can be executed over NYTimes; while t1 is executed over SWDF and Geonames, and t2 on all the three datasets. Thus, FedX produces a complete answer by joining the results obtained from executing these three subqueries. Nevertheless, FedX requires 239.4 s to execute the query.

ANAPSID offers two query decomposition methods: SSGS and SSGM (Fig. 2a). ANAPSID SSGS only selects one relevant source per triple pattern; execution time is reduced to 0.338 s, but sources are erroneously selected and the query execution produces empty results. Finally, ANAPSID SSGM builds a star-shaped subquery that includes t2 and t3. The star-shaped subquery is executed on NYTimes, while t1 is posed over SWDF and Geonames. Execution time is reduced, but only 19 answers are produced, i.e., results are incomplete.

Based on the values of join cardinality reported in Fig. 2b, the decomposition that produces all the results requires that t2 is executed over NYTimes and Geonames, while t1 and t3 should be only executed in Geonames and NYTimes, respectively. However, because of the lack of source description, neither FedX nor ANAPSID is capable of finding this decomposition. On the one hand, to ensure completeness, FedX selects irrelevant sources for t1 and t2, negatively impacting execution time. On the other hand, ANAPSID SSGS blindly prunes the relevant sources for t1 and t2, and does not collect data from Geonames and NYTimes required to answer the query. Similarly, ANAPSID SSGM prunes Geonames from t2, while it is unable to decide irrelevancy of Geonames in t1.

MULDER describes sources with RDF molecule templates (RDF-MTs), and will be able to select the RDF-MTs of the correct relevant sources. Thus, MULDER will produce complete answers without impacting on execution time.

1.2   Background

In this section, we present basic concepts required to understand this work. The Resource Description Framework (RDF) is a graph-based data model where nodes (subjects and objects) are connected via directed edges (predicates)[1]. Nodes in RDF graph can be resources or literals, and RDF resources are identified by IRIs (Internationalized Resource Identifier) or blank nodes (anonymous resources or existential variables). Literals used for string, number, and date values. An RDF triple consists of a subject, a predicate, and an object. A set of RDF triples are called RDF graph, and a collection of RDF graphs form an RDF dataset. In this paper, we use the term dataset and data source interchangeably. RDF resources can be served via native web access interfaces such as dereferencing resource identifiers, and web application programming interfaces (APIs) using SPARQL endpoint protocol.

Definition 1 (RDF Triple and Dataset [31]). *Let U, B, L be disjoint infinite sets of URIs, blank nodes, and literals, respectively. A tuple $(s, p, o) \in (U \cup B)\ X\ (U)\ X\ (U \cup B \cup L)$ is denominated an RDF triple, where s is called the subject, p the predicate, and o the object. An RDF dataset or RDF graph is a set of RDF triples. When $s \in L$ or $p \in (B \cup L)$, then the tuple (s, p, o) is called a generalized RDF triple and the dataset where it is contained is called a generalized RDF dataset [20].*

A set of triples that share same subject value are called *RDF molecules*. Formally, RDF molecules are defined as follows:

Definition 2 (RDF Molecule [12]). *Given an RDF graph G, an RDF molecule $\mathbb{M} \subseteq G$ is a set of triples $\mathbb{M} = \{t_1, t_2, \ldots, t_n\}$ in which $subject(t_1) = subject(t_2) = \cdots = subject(t_n)$.*

SPARQL query language is a W3C recommendation for querying RDF data. The SPARQL query language provides four query forms[2]: ASK, SELECT, CONSTRUCT, and DESCRIBE. In this work, we focus on SPARQL SELECT queries formally defined as follows:

Definition 3 (SPARQL Expression (SELECT Query) [39]). *Let V be a set of variables disjoint from $U \cup B \cup L$. A SPARQL expression is built recursively as follows. (1) A triple pattern $t \in (U \cup V)\ X\ (U \cup V)\ X\ (L \cup U \cup V)$ is an expression. (2) If $Q1$ and $Q2$ are expressions and R is a filter condition, then $Q1$ Filter R, $Q1$ Union $Q2$, $Q1$ Opt $Q2$, $Q1$ AND $Q2$ are expressions. Let Q be a SPARQL expression and $S \subset V$ a finite set of variables. A SPARQL SELECT query is an expression of the form $SELECT S_s(Q)$.*

SPARQL language is based on matching graph patterns; a basic graph pattern (BGP) is a set of triple patterns and filter patterns.

Definition 4 (BGP). *Let I be the set of all IRIs, B be the set of blank nodes, L be the set of literals and V be the set of variables. A SPARQL basic graph pattern (BGP) expression is defined recursively as follows:*

---

[1] https://www.w3.org/TR/rdf11-concepts/.

[2] https://www.w3.org/TR/2013/REC-sparql11-query-20130321/.

1. *A triple pattern* $tp \in (I \cup B \cup V) \times (I \cup V) \times (I \cup B \cup L \cup V)$ *is a BGP;*
2. *The expression (P1 AND P2) is a BGP, where P1 and P2 are BGPs;*
3. *The expression (P FILTER E) is a BGP, where P is a BGP and E is a SPARQL filter expression that evaluates to Boolean value.*

A BGP in a SPARQL query contains at least one star-shaped subquery (SSQ). An SSQ is a non-empty set of triple patterns that share the same subject variable (constant).

Definition 5 (Star-shaped Subquery (SSQ) [42]**).** *A star-shaped subquery* star(S,?X) *on a variable (constant) ?X is defined as:*

1. star(S,?X) *is a triple pattern* t = {?X p o}, *and* p *and* o *are different to ?X.*
2. star(S,?X) *is the union of two stars,* star(S1,?X) *and* star(S2,?X), *where triple patterns in* S1 *and* S2 *only share the variable (constant) ?X.*

SPARQL queries are evaluated over an RDF dataset based on mappings, where each mapping represents a possible answer of a query.

Definition 6 (SPARQL Mappings [39]**).** *A mapping is a partial function* $\mu : V \to (B \cup L \cup U)$ *from a subset of variables to RDF terms. The domain of a mapping* μ, $dom(\mu)$, *is the subset of V for which* μ *is defined. Two mappings* μ_1, μ_2 *are compatible, written* $\mu_1 \sim \mu_2$, *if* $\mu_1(x) = \mu_2(x)$ *for all* $x \in dom(\mu_1) \cap dom(\mu_2)$. *Further,* $vars(t)$ *denotes all variables in triple pattern t, and* $\mu(t)$ *is the triple pattern obtained when replacing all* $x \in dom(\mu) \cap vars(t)$ *in t by* $\mu(x)$.

A federated query processing engine provides a unified access interface for federation of RDF data sources accessible via Web interfaces. While distributed RDF storage systems have control over each dataset, federated query engines have no control over datasets in the federation and data providers can join or leave the federation at any time and modify their datasets independently. Saleem et al. [32] study federated query engines with Web access interfaces. Based on their survey results, the authors divide federation approaches into three main categories: *Query Federation over SPARQL endpoints, Query Federation over Linked Data (via URI lookups),* and *Query Federation on top of Distributed Hash Tables*; in this work, we focus on *Query Federation over SPARQL endpoints.*

2 Related Work

2.1 Federated Query Engines

The problem of query processing over federations of data sources has been extensively studied by the database and semantic web communities. Existing solutions rely on a global or unified interface that allows for executing queries on a federation of autonomous, distributed, and potentially heterogeneous data sources in a way that execution time is minimized while query completeness is maximized. In general, federated query processing systems have three main challenges: source selection, query decomposition, and query execution. Database federated engines employ the relational model to represent the unified view of the federation [14,17,18,22,44], and the query language SQL is utilized to express queries against the federation of data sources. Semantic federated query engines exploit the semantics encoded into the RDF datasets of a federation to build a catalog of source descriptions [2,5,7,15,25,42]. The catalog is used to select the sources from the federation where SPARQL queries will be executed. Existing semantic federated query engines include ANAPSID [2], FedX [40], Avalanche [5], Lusail [1], SPLENDID [15], and Semagrow [7]. Catalogs can be collected offline by the federated query engine, e.g., ANAPSID, or during query running-time, e.g., FedX. Furthermore, heuristic- or cost-based approaches use these catalogs to perform source selection and query decomposition. ANAPSID and FedX employ information about the RDF properties in the datasets of a federation to determine the SPARQL endpoints that can answer a SPARQL triple pattern. On the other hand, Avalanche, Lusail, SPLENDID, Odyssey, and Semagrow utilize statistics and cost models to select a query relevant data sources and fine low-cost query plans. However, both types of approaches may lead to selection of irrelevant sources that will not contribute to the final answer. MULDER combines both information about properties in each RDF molecule with shared RDF class and information about the links between them using RDF Molecule Templates (RDF-MTs), which are collected offline.

2.2 Describing and Selecting Data Source

Selecting the relevant data sources for a given query is one of the challenges in federated query processing. Identifying the relevant sources of a query not only leads to a complete answer but also faster execution time. SPARQL federation approaches can be divided into three categories [32]: *catalog-assisted (pre-computed), catalog-free (computed on-the-fly), and hybrid (uses both pre-computed and on-the-fly metadata)* solutions.

ANAPSID [2] collects information about the RDF predicates of the triple patterns that can be answered by an RDF dataset. During the source selection step, ANAPSID parses the SPARQL query into star-shaped subqueries and identifies the SPARQL endpoints for each subquery by utilizing predicate metadata. In contrast to ANAPSID, FedX [40] does not require metadata about the sources beforehand, but uses triple pattern-wise ASK queries. HiBISCuS [34] is a source selection method that uses a hybrid approach to collect dataset metadata; it combines capability metadata, which relies on authority fragment of URIs gathered for each endpoint, with triple-pattern wise ASK queries. HiBISCuS discards irrelevant sources for a particular query by modeling SPARQL queries as hypergraphs. Lusail [1], like FedX, uses a catalog-free solution for source selection and decomposition. Unlike FedX, Lusail takes an additional step to check if pairs of triple patterns can be evaluated as one subquery over a specific endpoint; this knowledge is exploited by Lusail during query decomposition and optimization.

SPLENDID [15] relies on instance-level metadata available as *Vocabulary of Interlinked Datasets* (VoID) [4] for describing the sources in a federation. SPLENDID provides a hybrid solution by combining VoID descriptions for data source selection along with SPARQL ASK queries submitted to each dataset at run-time for verification. Statistical information for each predicate and types in the dataset are organized as inverted indices, which will be used for data source selection and join order optimization. Similarly, Semagrow [7] implements a hybrid method like SPLENDID, pattern-wise source selection method which uses VoID descriptions (if available) and ASK queries. Although VoID allows for the description of a dataset statistics, this description is limited and lacks details necessary for efficient query optimization. For instance, though VoID descriptions provide information about link existence between datasets via a linking property, it is not clear in which class(es) this property belongs too. In addition, VoID descriptions could be out-of-date if the dataset updates are very frequent. Odyssey [25] collects detailed statistics information on datasets that enable cost estimation which may lead to low-cost execution plans. The optimization is based on a cost model using statistical methods used for centralized triple stores, i.e., Characteristics Set (CS) [29] and Characteristics Pairs (CP) [16,29]. Odyssey identifies CSs and sources using predicates of each star-shaped subquery. Then, it prunes to non-relevant sources based on links between star-shaped subqueries and by finding *Federated Characteristics Pairs (FCPs)*. However, unexpected changes and misestimated statistics may conduce to poor query performance.

Different data sources in a federation could contain duplicated data or can be replicas of a dataset. DAW [33] is a duplication-aware hybrid approach for triple pattern wise source selection; it uses the DAW index to identify sources that lead to duplicated results and skip those sources. After making triple pattern-wise source selection, the selected sources are ranked based on the number of new triples they provide; those sources that are below a threshold are skipped. Duplicates are detected using Min-Wise Independent Permutations (MIPs) stored in the DAW index for each triple within the same predicate. FEDRA [26] is a source selection strategy for sources with a high replication degree. FEDRA relies on schema-level fragment definitions and fragment containment to detect replication; it exploits replication information to minimize data redundancy and data transfer by reducing the number of unions, i.e., by minimizing the number of endpoints selected. Finally, LILAC [27] is a query decomposition technique that tackles data replication across data sources and SPARQL endpoints. MULDER RDF-MTs leverage semantics encoded in RDF sources and create logical partitions without any restrictions on the replication of the properties.

Wylot and Cudré-Mauroux [43] also utilize the notion of RDF molecule templates but provide only an intuitive description without formalization. On the contrary, we formally define RDF-MTs, and devise techniques for exploiting RDF-MTs during source selection, query decomposition, and planning. Unlike FedX and Lusail, MULDER collects RDF molecule templates (RDF-MTs) beforehand, reducing the number of requests sent to a data source during query time. MULDER describes sources as a set of RDF-MTs, where each RDF-MT describes an RDF class and its possible properties. An RDF-MT also contains a set of links that exist with in the same source and a set of links with other

RDF-MTs in different data sources. Given a SPARQL query MULDER parses it into star-shaped subqueries and creates a *query-graph* where nodes are star-shaped subqueries and edges are join variables. Using RDF-MT based source description, for each node in the *query-graph*, MULDER selects the RDF-MT(s) that contain all predicates of a star-shaped subquery. Finally, MULDER selects a source for a star-shaped subquery if it is described by a RDF-MT with properties that appear in the triple patterns of the subquery. Once the RDF-MT(s) are selected for the subqueries of a query, MULDER uses information about links between RDF-MT(s) to prune the RDF-MT(s) and select only the relevant sources; thus, speeding execution time without impacting query completeness.

2.3 Query Decomposition Techniques

Once sources are selected, subqueries are decomposed into a form that will be sent to each selected source. FedX introduces the concept of Exclusive Groups (EG) that combines a set of triple patterns that can be sent to same source. Lusail [1] presents a locality aware decomposition that uses the concept of *Global Join Variable* and source information to decompose queries, similar to Exclusive Groups in FedX. Semagrow [7] follows heuristics that group multiple triple patterns that could be sent to the same source into a single query. These heuristics utilize carnality estimations of combined triple patterns that reduce the search space of Semagrow optimizer. This optimization impacts on the optimization time. If data sources are known to mirror another source, then alternative plans are created, rather than a single plan. Odyssey [25] combines star-shaped subqueries to a single SPARQL query to a particular endpoint whenever the same source for star-shaped subqueries is selected. ANAPSID provides two heuristics for query decomposition [28]: SSGS (Star-Shaped Group Single endpoint selection) and SSGM (Star-Shaped Group Multiple endpoint selection). SSGS reduces the number of unions by selecting a source among relevant sources that can answer a star-shaped subquery. On the other hand, SSGM creates a set of UNIONs for each subquery that have more than one selected source. Although SSGS performs better in terms of execution time than SSGM, it could return incomplete results since it only selects one source per subquery. MULDER utilizes RDF-MTs in both source selection and query decomposition stages, reducing, thus, technological granularity and providing the semantics for enhancing federated query processing. MULDER combines only triple patterns in a star-shaped subquery as a single SPARQL query to be sent to a single source; thus, less source connections are required and query execution is sped up.

2.4 Query Optimization and Execution

Federated query engines, such as ANAPSID [2] and FedX [40], follow heuristics-based query optimization, while SPLENDID [15], Odyssey [25], and Semagrow [7] implement cost-based query optimization techniques to find a low cost plan. ANAPSID [2] is an adaptive federated query engine capable of delivering query results as soon as they arrive from the data sources, i.e., by implementing non-blocking join operators. On the other hand, FedX [40] is a non-adaptive federated query processing approach, at the level of query execution, that optimizes query execution by introducing exclusive groups, i.e., triple patterns that can be executed against only one source. The cost estimation of Semagrow [7] is based on a cost model over each operator using either statistics provided by source metadata or estimated cardinality of sub-expressions. The cost of operators is estimated by applying a communications overhead factor to the cardinality of the results. The cost of complex expressions is estimated recursively using a cost model over statistics about sub-expressions as well as distinct subjects and objects appearing in these results. Dynamic programming is used to enumerate different plans in order to identify the optimal one with respect to the cost model. Semagrow operates in an asynchronous and non-blocking way where operators subscribe to a stream and are notified when data becomes available. Similarly, Lusail [1] optimizes query execution using subquery ordering based on cardinality estimation on subqueries and projection list. Cost estimation is based on statistics collected at run-time on each triple pattern during query analysis. The generated decompositions lead to a set of subqueries with minimal execution cost. Using cardinality information of individual triple patterns, Lusail estimates the cardinality of subqueries and projection list. In addition, Lusail optimizes query execution by parallelism via process scheduling.

Odyssey [25] minimizes the number of subqueries that are posed to a source by combining subqueries that can be evaluated over exactly the same sources. First, it identifies an ordering of the triple patterns within each star-shaped subquery using Characteristic Set statistics. Then, cardinality of each subquery is estimated and a dynamic programming based algorithm is applied to identify a query plan. The cost function is defined based on the cardinalities of intermediate results and on how many results need to be transferred from sources during execution. Although in an ideal scenario, Odyssey may identify efficient query plans, collecting these detailed statistics is nearly impossible in a federated scenario where datasets are autonomous.

MULDER utilizes the connectivity between RDF-MTs to both prune the RDF-MTs of a query and identify the relevant sources associated with selected RDF-MTs. Furthermore, MULDER implements a heuristics-based that follows a Greedy strategy to identify a join ordering the subqueries. In consequence, query execution time is reduced and query answer completeness is enhanced.

3   Problem Statement and Proposed Solution

3.1   Problem Statement

In this section, we formalize the query decomposition and execution problems over a federation of RDF data sources.

Definition 7 (Query Decomposition). *Given a basic graph pattern BGP of triple patterns $\{t_1,\ldots,t_n\}$ and RDF datasets $D = \{D_1,\ldots,D_m\}$, a decomposition P of BGP in $D, \gamma(P|BGP,D)$, is a set of service graph patterns*

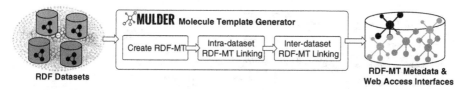

Fig. 3. Pipeline for creating RDF molecule templates (RDF-MTs). *Create RDF-MT:* Queries are executed over RDF datasets for creating RDF-MTs. *Intra-dataset RDF-MT Linking:* RDF-MTs are linked to other RDF-MTs in the same dataset. *Inter-dataset RDF-MT Linking:* RDF-MTs are associated with RDF-MTs in other datasets. Web interfaces are created to access RDF-MTs.

$SGP = (SQ, SD)$, *where SQ is a subset of triple patterns in BGP and SD is a subset of D.*

Definition 8 (Query Execution over a Decomposition). *The evaluation of $\gamma(P|BGP, D)$ in D, $[[\gamma(P|BGP, D)]]_D$, is defined as the join of the results of evaluating SQ over RDF datasets D_i in SD:*

$$[[\gamma(P|BGP, D)]]_D = JOIN_{(SQ,SD)\in\gamma(P|BGP,D)}(UNION_{D_i\in SD}[[SQ]]_{D_i}) \quad (1)$$

After we defined what a decomposition of a query is and how such a decomposed query can be evaluated, we can define the problem of finding a suitable decomposition for a query and a given set of data sources.

Definition 9 (Query Decomposition Problem). *Given a SPARQL query Q and RDF datasets $D = \{D_1, \ldots, D_m\}$, the problem of decomposing Q in D is defined as follows. For all BGPs, $BGP = \{t_1, \ldots, t_n\}$ in Q, find a query decomposition $\gamma(P|BGP, D)$ that satisfies the following conditions:*

- *The evaluation of $\gamma(P|BGP, D)$ in D is complete, i.e., if D^* represents the union of RDF datasets in D, then the results of evaluating BGP in D^* and the results of evaluating decomposition $\gamma(P|BGP, D)$ in D are the same, i.e.,*

$$[[BGP]]_{D^*} = [[\gamma(P|BGP, D)]]_D \quad (2)$$

- *$\gamma(P|BGP, D)$ has the minimal execution cost, i.e., if $cost(\gamma(P'|BGP, D))$ represents the execution time of a decomposition P' of BGP in D, then*

$$\gamma(P|BGP, D) = \underset{\gamma(P'|BGP,D)}{\text{argmin}} \; cost(\gamma(P'|BGP, D)) \quad (3)$$

3.2 Proposed Solution

To solve the query decomposing problem, we devise MULDER, a federated query engine for RDF datasets accessible through Web access interfaces, e.g., SPARQL endpoints. The MULDER *Decomposition & Source Selection* creates a query

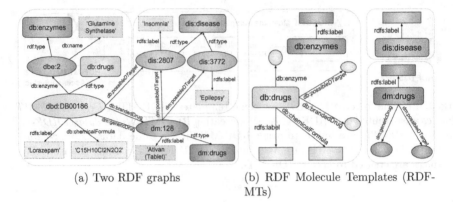

(a) Two RDF graphs   (b) RDF Molecule Templates (RDF-
                                        MTs)

Fig. 4. RDF-MT creation. Three RDF graphs with four RDF molecules of types:
db:drugs, db:enzymes, dm:drugs, and dis:disease, where prefixes *db* is for *drugbank*, *dm* is
for *dailymed*, and *dis* is for *diseasome* datasets. Four RDF Molecule Templates (RDF-
MTs) are created for these RDF classes.

decomposition with service graph patterns (SGPs) of star-shaped subqueries
built according to RDF-MT metadata. RDF-MTs describe the properties of the
RDF molecules contained in the RDF datasets, where an RDF molecule is a set
of RDF triples that share the same subject. Once the star-shaped subqueries
are identified, a bushy plan is built by the MULDER *Query Planning*; the plan
leaves correspond to star-shaped subqueries.

4   MULDER: A Federated Query Processing Engine

4.1   RDF Molecule Templates for Source Description

Web interfaces provide access to RDF datasets, and can be described in terms
of resources and properties in the datasets. MULDER relies on RDF-MTs to
describe the set of properties that are associated with subjects of the same
type in a given RDF dataset, i.e., the set of properties of the RDF molecules
contained in the dataset. In addition, MULDER extracts links between RDF-
MTs from the dataset schema and discovers unknown links from instances (intra-
and inter-links). Further, an RDF-MT is associated with a Web access interface
that allows for accessing the RDF molecules that respect the RDF-MT. Figure 3
presents the pipeline for RDF-MT creation.

Definition 10 (RDF Molecule Template (RDF-MT)). *An RDF Molecule
Template (RDF-MT) is a 5-tuple = < WebI, C, DTP, IntraL, InterL>, where:*

- *WebI – is a Web service API that provides access to an RDF dataset G via
  SPARQL protocol;*
- *C – is an RDF class such that the triple pattern (?s rdf:type C) is true in G;*

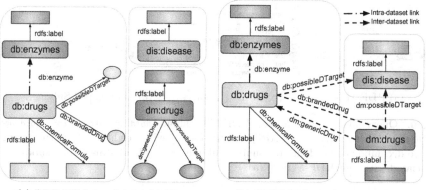

(a) RDF-MT Intra-dataset Linking   (b) RDF-MT Inter-dataset Linking

Fig. 5. RDF-MT linking. (a) Each RDF-MT is linked to other RDF-MTs in the same RDF dataset: db:drugs and db:enzymes. (b) Each RDF-MT is linked to other RDF-MTs in a different RDF dataset: db:possibleDiseaseTarget, dm:possibleDiseaseTarget, db:brandedDrug, and dm:genericDrug. Prefix *db* is for *drugbank*, *dm* is for *dailymed*, and *dis* is for *diseasome* datasets.

- *DTP – is a set of pairs (p, T) such that p is a property with domain C and range T, and the triple patterns (?s p ?o), and (?s rdf:type C) are true in G;*
- *IntraL – is a set of pairs (p, C_j) such that p is an object property with domain C and range C_j, and the triple patterns (?s p ?o) and (?o rdf:type C_j) and (?s rdf:type C) are true in G;*
- *InterL – is a set of triples (p, C_k, SW) such that p is an object property with domain C and range C_k; SW is a Web service API that provides access to an RDF dataset K, the triple patterns (?s p ?o) and (?s rdf:type C) are true in G, and the triple pattern (?o rdf:type C_k) is true in K.*

Algorithm 2 identifies the RDF Molecule Templates that describe the sources of a federation. Algorithm 2 first collects a list of RDF-MTs with their intra-dataset links for each SPARQL endpoint by calling Algorithm 1 (Line 3–5).

Algorithm 1. Collect RDF Molecule Templates: *ws*: a SPARQL endpoint

---

1: **procedure** COLLECTRDFMOLECULETEMPLATES(ws)
2:    $MTL \leftarrow [\ ]$ ▷ MTL - list of molecule templates
3:    $CP \leftarrow getClassesWithProperties(ws)$ ▷ SELECT query
4:    **for** $(C_i, \mathbb{P}_i) \in CP$ **do**   ▷ C_i - class axioms
5:       $\mathbb{L}_i \leftarrow \mathbb{L}_i + (p_i, C_j) | \exists p_i \in \mathbb{P}_i, (range(p_i) = C_j) \wedge ((C_j, \mathbb{P}_j) \in CP)$
6:       $RDF\text{-}MT_i \leftarrow (C_i, \mathbb{P}_i, \mathbb{L}_i)$
7:       $MTL \leftarrow MTL + RDF\text{-}MT_i$
8:    **end for**
9:    **return** MTL
10: **end procedure**

(a) RDF-MTs   (b) SPARQL Query   (c) Star-shaped Subqueries

Fig. 6. Query decomposition. (a) RDF-MTs about `db:drug_interaction`, `db:drugs`, `db:target`, `db:reference`. (b) SPARQL query composed of eight triple patterns that can be decomposed into four star-shaped subqueries. (c) Four star-shaped subqueries associated with four RDF-MTs in (a).

Algorithm 1 extracts RDF-MTs by executing SPARQL queries against a Web access interface ws of an RDF dataset, i.e., a SPARQL endpoint. First, *RDF classes* are collected with their corresponding *properties* (Line 3), i.e., pairs $(C_i, \mathbb{P}_i)$, where C_i is a class and $\mathbb{P}_i$ is a set of predicates of C_i. Figure 4 illustrates the creation of four RDF-MTs from two given RDF graphs (Fig. 4a). Then, for each RDF class C_i, object properties in $\mathbb{P}_i$ are identified, and the set of intra-dataset links (IntraL) are generated (Line 4–8). Figure 5a shows the result of generating intra-dataset links among three RDF-MTs in the same dataset. Then Algorithm 2 iterates over each RDF-MT in each Web access interface (WAI) and finds inter-dataset links between them (InterL) (Line 6–10). Finally, RDF-MTs are stored to a file (Line 11). Figure 5b presents inter-dataset links between RDF-MTs from two RDF graphs presented in Fig. 4a.

Algorithm 2. Create RDF Molecule Templates: WI: Set of SPARQL endpoints, and WAI: hash map of SPARQL endpoints to RDF Molecule Templates

---

```
 1: procedure CREATEMOLECULETEMPLATES(WI)
 2:     WAI ← {}                              ▷ WAI - a map of ws and MTLs
 3:     for ws_i ∈ WI do
 4:         WAI(ws_i) ← CollectRDFMoleculeTemplates(ws_i)      ▷ Algorithm 1
 5:     end for
 6:     for (ws_i, [(C_i, P_i, L_i)]) ∈ WAI do
 7:         for (ws_k, [(C_k, P_k, L_k)]) ∈ WAI and ws_k ≠ ws_i do
 8:             L_i ← L_i + (p_i, C_k) such that ∃p_i ∈ P_i ∧ range(p_i) = C_k
 9:         end for
10:     end for
11:     saveRDFMT(WAI)
12: end procedure
```

4.2 MULDER Source Selection and Query Decomposition Techniques

Given a SPARQL query MULDER parses the query into star-shaped subqueries and create a *query-graph* where nodes are star-shaped subqueries and edges

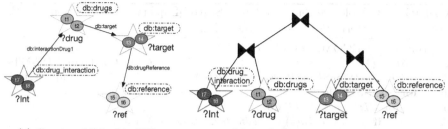

(a) Graphs of Joinable SSQs   (b) Bushy Plan of Joinable SSQs

Fig. 7. Query planning. (a) Joinable graph of star-shaped subqueries (SSQs) represents joins between SSQs. (b) Bushy plan of joinable star-shaped subqueries (SSQs). Graph of joinable SSQs is utilized by MULDER Query Planner to create a bushy plan of SSQs where joins between SSQs are maximized.

are join variables. Using RDF-MT based source description, for each node in the *query-graph*, MULDER selects RDF-MT(s) that contain all predicates of a star-shaped subquery. Finally, MULDER selects a source(s) for each star-shaped subquery, if the source contains an RDF-MT(s) with matching properties in a star-shaped subquery, MULDER applies pruning using the actual links that are known between RDF-MTs. MULDER combines triple patterns in a star-shaped subquery as a single SPARQL query to be sent to a single source.

Figure 6 shows an example of query decomposition and source selection. The example query in Fig. 6b, contains eight triple patterns. The first step of query decomposition is to identify the star-shaped subqueries (SSQ). In our example, four subqueries which contain two triples patterns each, are identified, i.e., ?drug (t_1, t_2), ?target (t_3, t_4), ?ref (t_5, t_6), and ?Int (t_7, t_8). Each of SSQs are then associated with RDF-MTs that contain predicates in SSQs, as shown in Fig. 6c.

The MULDER query decomposer is sketched in Algorithm 3. Given a BGP and a set of RDF-MTs (WIT), SSQs are first identified (Line 3). Then, RDF-MTs which contain all predicates in SSQ are determined from WIT as candidate RDF-MTs (Line 4–10). Furthermore, linked candidate RDF-MTs with respect to Joinable SSQs (Line 11) are identified (Line 12). Finally, candidate RDF-MTs are pruned, i.e., candidate RDF-MTs that contain all predicates in SSQ but are not linked to any RDF-MT that matches Joinable SSQ are excluded (Line 13). SSQs that have more than one matching RDF-MT from the same Web access interface will be decomposed as one service graph pattern. However, if matching RDF-MTs are from different Web access interfaces, then MULDER decomposes them; the UNION operator is used during query execution to collected the data from each Web access interface.

4.3 MULDER Query Optimization Strategies

Figure 7a shows joinable star-shaped subqueries (SSQ) that share at least one variable, i.e., ?Int is joinable with ?drug via predicate db:interactionDrug1, while ?drug is joinable with ?target via db:target. Furthermore, ?target is joinable

Algorithm 3. Molecule template based SPARQL query decomposition: BGP: Basic Graph Pattern, WIT: set of RDF-MTs

1: **procedure** DECOMPOSE(BGP, WIT)
2:    $CM \leftarrow \{\}$ ▷ CM - Candidate RDF-MTs
3:    $SSQs = getStarShapedSubqueries(BGP)$ ▷ Subject stars
4:    **for** $s \in SSQs$ **do**
5:        **for** $RDF_{MT} \in WIT$ **do**
6:            **if** $predicatesIn(s) \subseteq predicatesIn(RDF_{MT})$ **then**
7:                $CM[s].append(RDF_{MT})$
8:            **end if**
9:        **end for**
10:    **end for**
11:    $JSSQ = getJSSQs(SSQs)$ ▷ $Query\text{-}graph$ of Joinable SSQs
12:    $conn = connectedRDFMTs(SSQs, JSSQ, CM)$ ▷ selected RDF-MTs graph
13:    $DQ = prune(SSQs, JSSQ, conn)$
14:    **return** DQ ▷ decomposed query
15: **end procedure**

with ?ref through db:drugReference property. Finally, MULDER query planner generates bushy plans combining SSQs (Fig. 7b). The problem of identifying a bushy plan from conjunctive queries is known to be NP-complete [35]. MULDER planner implements a greedy heuristics based approach to generate a bushy plan, where the leaves correspond to SSQs, and the number of joins between SSQs is maximized while the plan height is minimized.

4.4 The MULDER Architecture

The MULDER architecture is depicted in Fig. 8. The MULDER *Query Processing Client* receives a SPARQL query, decomposes it, performs source selection based on RDF-MT metadata, and, afterwards, identifies a bushy plan against the selected RDF datasets. The MULDER *Query Engine* executes the bushy plan and contacts the MULDER query processing server to evaluate Service Graph Patterns over the Web access interfaces. Further, the MULDER *Query Processing Server* receives requests from the MULDER client to retrieve RDF-MT metadata about RDF datasets, e.g., metadata about properties of RDF molecules contained in these RDF datasets.

5 Experimental Study

We empirically study the efficiency and effectiveness of MULDER query decomposition and source selection techniques. We assess the query performance of MULDER utilizing RDF molecule templates and templates generated using the METIS [24] and SemEP [30] partitioning algorithms. Furthermore, we compare MULDER with the federated query engines ANAPSID and FedX for three well-established benchmarks – BSBM, FedBench, and LSLOD. Finally, we evaluate

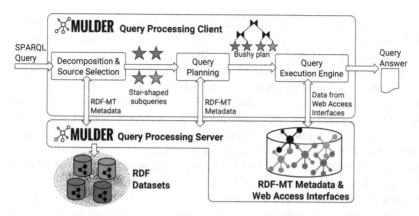

Fig. 8. The MULDER client-server architecture. MULDER query processing client receives SPARQL queries, creates query decompositions with star-shaped subqueries, and identifies and executes bushy plans. MULDER query processing server collects both RDF-MT metadata about RDF datasets and results of executing queries over Web access interfaces, e.g., SPARQL endpoints.

the *continuous efficiency* of MULDER compared to ANAPSID. The following research questions are evaluated: **(RQ1)** Do RDF-MTs characterize the semantics represented within and between data sources? **(RQ2)** Do different MULDER source descriptions impact on query processing in terms of efficiency and effectiveness? **(RQ3)** Is the effectiveness and efficiency of the query processing process impacted by the MULDER query decomposition technique? **(RQ4)** Is the continuous efficiency of the answer generation process impacted by the MULDER query decomposition technique?

The MULDER[3] decomposition and source selection, and query planning components are implemented in Python 2.7.10, and integrated into ANAPSID [2], i.e., MULDER plans are executed using ANAPSID physical operators. Experiments are executed on two Dell PowerEdge R805 servers, AMD Opteron 2.4 GHz CPU, 32 cores, 256 GB RAM. BSBM (The Berlin SPARQL Benchmark), FedBench, and LSLOD datasets are deployed on one machine as SPARQL endpoints using *Virtuoso 6.01.3127*, where each dataset resides in a dedicated Virtuoso docker container.

5.1 Benchmarks

Three benchmarks are utilized to assess our research questions: (i) BSBM - The Berlin SPARQL Benchmark; (ii) FedBench; and (iii) LSLOD. For each RDF dataset of the federations of these benchmarks, the RDF-MTs are computed; furthermore, graph analytics are utilized to described the properties of these RDF datasets in terms of the RDF-MTs and their links. We generated all RDF-MTs and their interconnections considering both intra-dataset and inter-dataset

---

[3] https://github.com/SDM-TIB/MulderTLDKS.

links, as defined in Algorithms 1 and 2, respectively. We use graph density, connected components, transitivity, and average clustering coefficient to analyze the main properties of the graph that models the RDF-MTs of each federation. Clustering coefficient measures the tendency of nodes who share same connections in a network to become connected. If the neighborhood is fully connected, the clustering coefficient is 1 and a value close to 0 means that there are no connections in the neighborhood. Average clustering coefficient assigns higher scores to low degree nodes, while the transitivity ratio places more weight on the high degree nodes. The average connectivity of a graph is the average of local node connectivity over all pairs of nodes of the graph. We model a graph of RDF-MTs as undirected a multi-graph (MultiGraph in networkx[4]) where the predicates that connect each RDF-MTs are used as labels of the edges. A multi-graph is used to compute the number of nodes, edges, average number of neighbors, connected components, and average node connectivity. Finally, we model the graph as undirected single network graph, where a link between RDF-MTs is represented as unlabelled edges. Using single network graphs, transitivity and average clustering coefficient are computed.

Table 1. BSBM queries characteristics. X SPARQL clause – refers to AND Join operation between triple patterns in a BGP.

Query	#BGPs	#TriplePatterns	#SSQs	SPARQL clauses
B1	1	4	1	DISTINCT
B2	1	6	1	X
B3	2	12	2	DISTINCT, UNION
B4	5	17	7	OPTIONAL
B5	1	8	2	DISTINCT
B6	2	10	4	OPTIONAL
B7	1	8	2	DISTINCT
B8	1	4	2	X
B9	1	5	2	DISTINCT
B10	1	11	3	DISTINCT
B11	1	25	4	DISTINCT
B12	1	21	4	DISTINCT

BSBM - The Berlin SPARQL Benchmark The Berlin SPARQL Benchmark [6] is a synthetic dataset focusing on an e-commerce use case where a set of products are offered by different vendors and consumers and reviewers have posted reviews about these products on various review sites. The data model contains

---

[4] https://networkx.github.io/.

eight classes: Product, ProductType, ProductFeature, Producer, Vendor, Offer, Review, and Person. The benchmark queries contain parameters (placeholders) which are enclosed with % characters. During an instantiation of the queries, these parameters are replaced with random values from the benchmark dataset. We use BSBM to generate 12 SELECT queries (with 20 instantiations each) over a generated dataset containing 200 million triples[5]; characteristics of these queries are presented in Table 1. We partitioned the dataset using *rdf:type* classes and created eight SPARQL endpoints, one per each class and one endpoint which contains the whole dataset.

The first benchmark for our experiment, and the smallest in number of RDF-MTs, is the Berlin SPARQL Benchmark (BSBM). In BSBM benchmark, there are only eight RDF-MTs and eight links between them. For our experiment we treated each RDF-MT as a single dataset by creating a separate endpoint for them. Therefore, there are in total eight datasets and since each dataset contains only one RDF-MT, there are no intra-dataset links for this benchmark. Figure 9 illustrates all RDF-MTs in BSBM where each contained in a single dataset (hence different colors) and their inter-dataset connections[6]. In addition, in order to study the characteristics of the generated BSBM molecule template graph, we report on a graph analysis in Table 2. We observed a strong connection between RDF-MTs - 0.285 density and 1.0 average node connectivity. In particular, the connections concentrated on a single RDF-MT (hence, a single dataset), Product, with 6 out of 8 links to or from this RDF-MT (hence, dataset). A histogram of frequencies of RDF-MTs per number of properties distributed from six (two RDF-MTs) to 18 (one RDF-MT) is shown in Fig. 10.

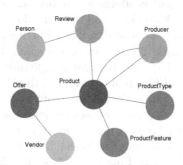

Fig. 9. Analysis of RDF-MTs of BSBM. The graph comprises 8 RDF-MTs and 8 inter-dataset links. Each dot represents an RDF-MT stored in each endpoint. A line between dots corresponds to inter-dataset links. There is only one RDF-MT in each endpoint, hence no intra-dataset links.

---

[5] BSBM queries can be found in the Appendix A.
[6] The graph visualization was generated using the open source software platform cytoscape – http://www.cytoscape.org/.

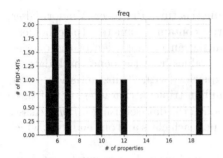

Fig. 10. Frequency of BSBM RDF-MTs per number of properties. Majority of molecule templates contain from five to seven properties.

Table 2. FedBench RDF-MT graph metrics. Clustering coefficient (0.0) suggests that there is no connectivity in the neighborhood of the network.

Num of nodes	8
Num of edges	8
Graph density	0.285
Avg. num of neighbors	2
Connected components	1
Avg. node connectivity	1.0
Transitivity	0.0
Clustering coefficient	0.0

FedBench

FedBench [36] is a benchmark suite for analyzing both the efficiency and effectiveness of federated query processing strategies for different use cases on semantic data. It includes three collections of datasets: **cross-domain**, **life-science**, and **SP^{2}Bench** collections. The **cross-domain** collection is composed of datasets from different domains: DBpedia has linked structured data extracted from Wikipedia; Geonames is composed of geo-spacial entities such as countries and cities; Jamendo includes music data such as artists, records; LinkedMDB maintains linked structured data about movies, actors; the New York Times dataset contains about 10,000 subject headings about people, organizations, and locations; finally, the Semantic Web Dog Food (SWDF) dataset includes data about Semantic Web conferences, papers, and authors. Furthermore, **Life-science** collection contains datasets from the life-sciences domain: Kyoto Encyclopedia of Genes and Genomes (KEGG) has chemical compounds and reactions data in Drug, Enzyme Reaction and Compound modules; the Chemical Entities of Biological Interest (ChEBI) contains information about molecular entities on "small" chemical compounds, such as atoms, molecules, ions; and DrugBank maintains drug data with drug target information. In addition to these three datasets in the life-sciences collection, a subset of DBpedia dataset that includes data about drugs is added in this collection. Finally, SP^2**Bench** collection contains a synthetic dataset generated by the SP^{2}Bench data generator [38], that mirrors characteristics observed in the DBLP database. For our experiments, we have used datasets from the first two collections from this benchmark, i.e., **cross-domain** and **life-science** collections, which contain real-world datasets.

We run 25 FedBench queries[7], including cross-domain queries (CD), linked data queries (LD), and life science queries (LS). Additionally, 10 complex queries (C) proposed by [42] are considered. The queries are executed against the Fed-Bench datasets from **cross-domain** and **life-science** collections. A SPARQL

---

[7] FedBench queries can be found in http://fedbench.fluidops.net/resource/Queries.

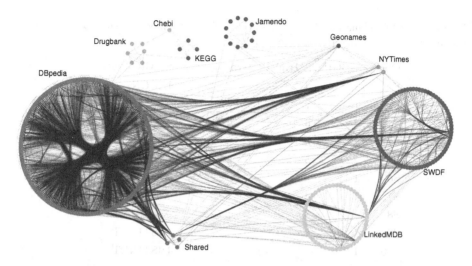

Fig. 11. Analysis of RDF-MTs of FedBench. The graph comprises 387 RDF-MTs and 6,317 intra- and inter-dataset links. The dots in each circle represent RDF-MTs. A line between dots in the same circle shows intra-dataset links, while a line between dots in different circles corresponds to inter-dataset links. In numbers, there is only one RDF-MT in ChEBI, 234 in DBpedia, six in Drugbank, one in Geonames, 11 in Jamendo, four in KEGG, 53 in LinkedMDB, two in NYTimes, and 80 in SWDF dataset. Four of these RDF-MTs belong to at least two FedBench datasets, modeled as separate circular dots.

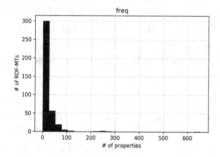

Fig. 12. Frequency of FedBench RDF-MTs per number of properties. Majority of molecule templates contain from one to 20 properties.

Table 3. FedBench RDF-MT graph metrics. Clustering coefficient (0.602) suggests high number of intra- & inter-dataset links.

Num of nodes	396
Num of edges	6,317
Graph density	0.081
Avg. num of neighbors	31.904
Connected components	9
Avg. node connectivity	10.624
Transitivity	0.395
Clustering coefficient	0.602

endpoint able to access a unified view of all the FedBench datasets (i.e., the RDF dataset D^* in Eq. 2) serves as gold standard and baseline.

In FedBench, RDF-MTs that have more than 100 properties correspond to classes with multiple predicates and subclasses, such as dbo:Person, dbo:Organisation, and dbo:Place. In addition, in order to study the characteris-

Table 4. LSLOD queries characteristics. X SPARQL clause – refers to AND Join operation between triple patterns in a BGP.

Query	#BGPs	#TriplePatterns	#SSQs	SPARQL clauses
S1	2	4	2	UNION
S2	1	7	3	X
S3	1	6	4	X
S4	1	5	2	DISTINCT
S5	2	5	3	OPTIONAL
S6	1	3	2	X
S7	1	4	3	X
S8	1	3	2	X
S9	1	8	2	DISTINCT
S10	1	8	2	DISTINCT

tics of the generated FedBench RDF-MT graph, we report on a graph analysis which is documented in Table 3. In particular, we observe a rather medium connectivity of the graph nodes (i.e., RDF-MTs) – 0.081 – with 31.9 average number of neighbors and 9 connected components[8]. Finally, the clustering coefficient (0.602) indicates that we do not have only links between the RDF-MTs that come from the same dataset, but also many inter-dataset connections. Figure 11 illustrates all RDF-MTs in FedBench associated with the dataset they come from with all intra-dataset and inter-dataset connections. In total, 387 RDF-MTs (396 including shared RDF-MTs) with 6, 317 links are generated. While the majority of the RDF-MTs (230) are related to a single dataset, quite a few (4) are shared between two or more datasets. Most of the RDF-MTs have between three and 20 properties, as can be seen in the histogram of Fig. 12.

LSLOD

LSLOD [19] is a benchmark composed of 10 real-world datasets of the Linked Open Data (LOD) cloud from life sciences domain. The federation includes: ChEBI (the Chemical Entities of Biological Interest), KEGG (Kyoto Encyclopedia of Genes and Genomes), DrugBank, TCGA-A (subset of The Cancer Genome Atlas), LinkedCT (Linked Clinical Trials), Sider (Side Effects Resource), Affymetrix, Diseasome, DailyMed, and Medicare. We run 10 simple queries[9] provided for LSLOD datasets in [19], characteristics of these queries are presented in Table 4. Compared to FedBench, LSLOD datasets contain rather small number of RDF-MTs. Figure 13 shows the connectivity of all RDF-MTs associated with LSLOD datasets. In total, there are 57 RDF-MTs with 197 links between them. TCGA-A dataset contains the majority of RDF-MTs (23). There are no shared RDF-MTs between the LSLOD datasets. Figure 14 shows that most of the

---

[8] A lower number of connected components indicates a stronger connectivity.

[9] LSLOD queries can be found in Appendix A.

RDF-MTs have between three and 55 properties. Some RDF-MTs from TCGA-A have a large number of properties, e.g., `tcga:clinical_omf` has 197 properties; `tcga:normal_control`, `tcga:tumor_sample`, and `tcga:clinical_nte` have 246 properties; and `tcga:clinical_cqcf`, `tcga:biospecimen_cqcf`, and `tcga:patient` have 247 properties. Graph analysis in Table 5 shows that there is medium connectivity (stronger than FedBench) of RDF-MTs, with 0.123 density, 6.912 average number of neighbors, and 3 connected components.

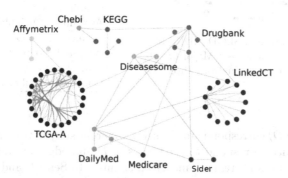

Fig. 13. Analysis of RDF-MTs of LSLOD. The graph comprises 56 RDF-MTs and 197 intra- and inter-dataset links; dots in each circle represent RDF-MTs. A line between dots in the same circle shows intra-dataset links, while a line between dots in different circles corresponds to an inter-dataset link. There are nine datasets: Drugbank, Dailymed, Sider, Affymetrix, KEGG, LinkedCT, TCGA-A, ChEBI, and Medicare; they have six, three, two, three, four, 13, 23, one, and one RDF-MTs, respectively.

From the reported analysis, it can be observed that RDF-MTs can be used to describe characteristics of datasets in terms of connectivity between RDF types represented in each dataset with other datasets in the federation. This answers **RQ1** positively in a sense that datasets can be characterized not only by ontology types (RDF types) and predicates, but also using the characteristics of the network between ontology types within the same dataset and with other datasets in a federation.

5.2 Comparison of MULDER Source Descriptions

We study the impact of MULDER RDF-MT on query processing, and compare the effect of computing molecule templates using two existing graph partitioning methods: METIS and SemEP. We name MULDER-SemEP and MULDER-METIS, the version of MULDER where molecule templates have been computed using SemEP and METIS, respectively. Co-occurrences of predicates in the RDF triples of a dataset D are computed. Given predicates p and q in D, co-occurrence of p and q ($co(p,q,D)$) is defined as follows:

$$co(p, q, D) = \frac{|subject(p, D) \cap subject(q, D)|}{|subject(p, D) \cup subject(q, D)|} \qquad (4)$$

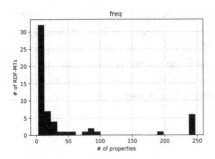

Fig. 14. Frequency of LSLOD RDF-MTs per number of properties. Majority of molecule templates contain from three to 30 properties

Table 5. LSLOD RDF-MT graph metrics. Clustering coefficient (0.375) suggests high number of intra- & inter-dataset links

Num of nodes	57
Num of edges	197
Graph density	0.205
Avg. num of neighbors	11.474
Connected components	3
Avg. node connectivity	1.648
Transitivity	0.634
Clustering coefficient	0.375

where $subject(q,D)$ corresponds to the set of different subjects of q in D. A graph GP_D where nodes correspond to predicates of D and edges are annotated with co-occurrence values is created, and given as input to SemEP and METIS. The number of communities determined by SemEP is used to create the corresponding partitions for METIS. METIS and SemEP molecule templates are composed of predicates with similar co-occurrence values. Each predicate is assigned to only one community. For this experiment, we use the following metrics: *(i) Execution Time*: Elapsed time between the submission of a query to an engine and the delivery of the answers. Time corresponds to absolute wall-clock system time as reported by the Python `time.time()` function. Timeout is set to 300 s. *(ii) Cardinality*: Number of answers returned by the query. Figure 15 reports on execution time and answer cardinality of the BSBM queries. The observed results suggest that knowledge encoded in RDF-MTs allows MULDER to identify query decompositions and plans that speed up query processing by up to two orders of magnitude, while answer completeness is not affected. Specifically, MULDER-RDF-MTs is able to place in SSQs non-selective triple patterns, while MULDER-SemEP and MULDER-METIS group non-selective triple patterns alone in subqueries. Thus, the size of intermediate results is larger in MULDER-SemEP and MULDER-METIS plans, impacting execution time. MULDER-RDF-MTs is able to provide a complete answers for all queries, while MULDER-METIS timed out for queries B5, B10, B11 and B12, and, respectively, MULDER-SemEP for queries B8 and B10. In terms of execution time, MULDER-RDF-MTs performs better than both MULDER-METIS and MULDER-SemEP on all queries. The observed results allow us to positively answer **RQ2**, and conclude that RDF-MTs based source descriptions improve the performance of query processing, compared to state-of-the-art graph partitioning methods.

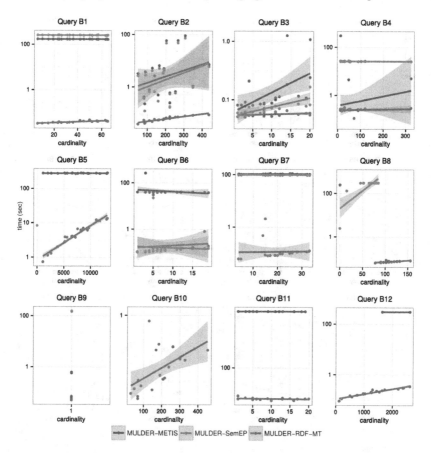

Fig. 15. BSBM: Performance for MULDER source descriptions. RDF molecules are computed using: Algorithm 1, SemEP, and METIS. RDF-MTs allow MULDER to identify query decompositions and plans that speed up query processing by up to two orders of magnitude, without affecting completeness.

5.3 Comparison of Federated Query Engines

We evaluate the efficiency and effectiveness (in terms of execution time and answer completeness, respectively) of RDF-MT query processing technique implemented in MULDER compared with the state-of-the-art federated query engines, FedX and ANAPSID. We created a unified view of all datasets in a benchmark via a *direct* SPARQL endpoint as a baseline. For this experiment, we compare federated query processing techniques using the following metrics: *(i) Execution Time*: Elapsed time between the submission of a query to an engine and the delivery of the answers. Time corresponds to absolute wall-clock system time as reported by the Python `time.time()` function. Timeout is set to 300 s. *(ii) Cardinality*: Number of answers returned by the query. *(iii) Completeness*:

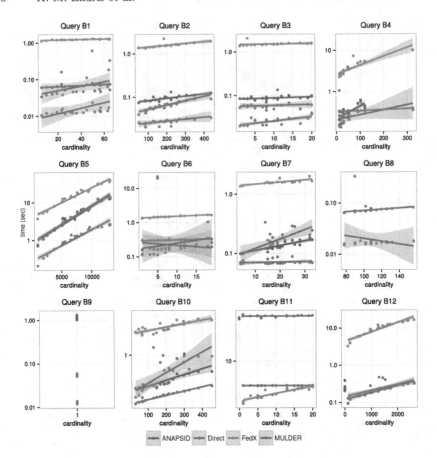

Fig. 16. BSBM: performance of federated engines. MULDER and ANAPSID outperform FedX in terms of query execution time, while MULDER overcomes ANAPSID in terms of completeness. Direct represents a unified SPARQL endpoint over one dataset with all the federation RDF triples.

Query result percentage with respect to the answers produced by the unified SPARQL endpoint created as the union of all datasets in the benchmark.

Performance of BSBM Queries. Figure 16 reports on the throughput of the federated engines ANAPSID, FedX, and MULDER for all BSBM queries. In many queries, MULDER and ANAPSID exhibit similar query execution times. FedX is slower than the two federated engines by at least one order of magnitude. ANAPSID returns query answers fast but at the cost of completeness, as can be observed in the queries B4, B7, B11, and B12. In addition, FedX and ANAPSID fail to answer B8 which is completely answered by MULDER.

Performance of FedBench Queries. Figure 17 visualizes the results of the four FedBench groups of queries (CD, LD, LS, C) in terms of answer completeness and query execution time. Measurements that are located in Quadrants I and III indicate bad performance and incomplete results, points in Quadrant IV are the best in terms of execution time and completeness, i.e., they correspond to a solution to the query decomposition problem; finally, points in Quadrant II show complete results but slower execution times. MULDER outperforms ANAPSID and FedX with regard to the number of queries it manages to answer: ANAPSID answers 29, FedX 27, and MULDER 31 out of 35 queries (Query C9 could not be answered by any of the engines). In particular, MULDER delivers answers to queries C1, C3, C4, LS4, LS5, and LS6 for which FedX fails and CD6 and LD6 for which ANAPSID fails. FedX returns complete and partially complete results for 20 and 7 queries respectively, exhibiting high execution times though (>1 s). In comparison to ANAPSID, MULDER achieves in general higher completeness of results, but at the cost of query execution time. For instance, C2, C8, and LD1 are answered by ANAPSID faster by almost one order of magnitude. Results observed in both benchmarks, i.e., BSBM and FedBench, allow us to positively answer **RQ2** and **RQ3**, and conclude that RDF-MTs enable MULDER decomposition and planning methods to identify efficient and effective query plans.

5.4   Measuring Continuous Efficiency of MULDER

In this experiment, we evaluate the efficiency of MULDER in terms of continuous generation of answers of a query. A continuous efficiency (diefficiency) of a query engine can be analyzed from the answer traces [3]. Answer traces are a sequence of pairs (t_i, μ_i) where t_i is the time-stamp that the ith answer, μ_i, is produced. There are two methods, recently proposed in [3], to measure the continuous effect of the engine, $dief@t$ and $dief@k$. Values of these metrics correspond to the number of answers produced in function of time, also known as *Answer Distribution Function*. Diefficiency at time t, $dief@t$, measures the continuous efficiency of an engine in the first t time units of query execution, while diefficiency at k answers, $dief@k$, measures the diefficiency of an engine while producing the first k answers of a query after the first answer is produced.

Dief@t metrics computes AUC (area-under-the-curve) of answer distribution until time t. Given an approach ρ, a query Q, and answer distribution function $X_{\rho,Q}$ while ρ executes Q, $dief@t$ is computed as:

$$dief_{\rho,Q}@t := \int_0^t X_{\rho,Q}(x)dx \tag{5}$$

Dief@k metrics computes AUC of answer distribution until the point in time t_k when the engine produces the kth answer, as recorded in the answer trace. Given an approach ρ, a query Q, and answer distribution function $X_{\rho,Q}$ while ρ executes Q, $dief@k$ is computed as:

$$dief_{\rho,Q}@k := \int_0^{t_k} X_{\rho,Q}(x)dx \tag{6}$$

where $t_k \in \mathbb{R}$ is the point in time when ρ produces the kth answer of Q.

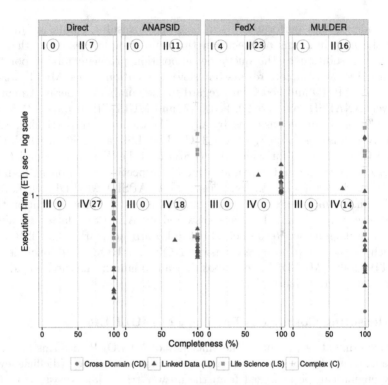

Fig. 17. FedBench: execution time and completeness of federated engines. Plots are divided into four quadrants: Incomplete results and slower execution time are reported in Quadrant I; results in Quadrant II correspond to complete results with lower performance; Quadrant III reports faster execution time but incomplete results; Quadrant IV indicates complete results and faster execution time. ANAPSID, FedX and MULDER manage to answer 29, 27, and 31 queries, respectively. Direct represents a unified SPARQL endpoint that is able to answer 34 of the 35 benchmark queries before timing out.

Since both ANAPSID and MULDER generate results incrementally, we compared the two heuristics techniques of ANAPSID, i.e., SSGS and SSGM, with MULDER using diefficiency metrics. Figure 18 shows answer traces and continuous query performance, of the three approaches for S2, S3, S4, S5, S7, and S8 LSLOD benchmark queries. As shown in Fig. 18, for these queries MULDER outperforms both ANAPSID heuristics techniques. On S2, the answer trace shows that MULDER produced more results faster in the first 0.05 secs of overall execution than other both ANAPSID approaches. All approaches return 0 results for query S8, none of the data sources in the benchmark is able to answer the query. Even though, the result is empty, MULDER produces the empty answer earlier than ANAPSID and without contacting any source. Therefore, we excluded query S8 from the reported results on the continuous efficiency of approaches in this section. Figure 19 shows answer traces for S1, S6, S9, and S10

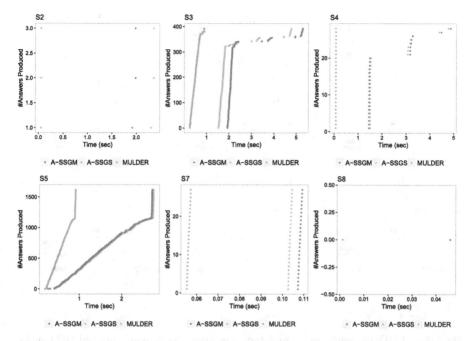

Fig. 18. LSLOD: answer traces. Continuous query performance. Y-axis shows the number of answers produced, and x-axis shows time in seconds (time.time()).

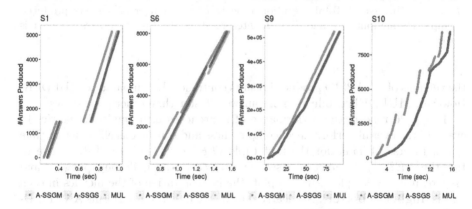

Fig. 19. LSLOD: answer traces. Continuous query performance. Y-axis shows the number of answers produced, and x-axis show time in seconds (time.time()).

LSLOD benchmark queries where all approaches produced answers in a uniform way. On these queries, MULDER produced answers faster than both ANAPSID approaches. For queries S1 and S6, MULDER is able to produce the first answer (row) earlier than ANAPSID-SSGS and ANAPSID-SSGM. For query S6 though, all approaches produce results continuously, MULDER produces slightly higher number of answers faster than others until the first sec. Finally, on query S10,

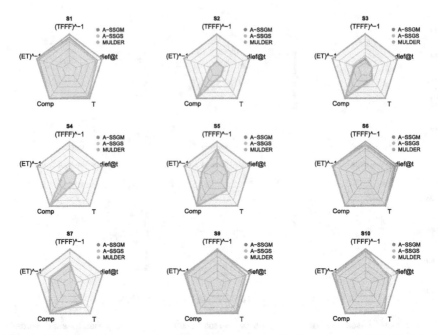

Fig. 20. LSLOD: efficiency and completeness metrics. Performance per LSLOD benchmark query of ANAPSID-SSGS (A-SSGS), ANAPSID-SSGM (A-SSGM) and MULDER query approaches. Axes correspond to: Inverse of Time for the first tuple $(TFFT^{-1})$, Inverse of Total execution time (ET^{-1}), Number of answers produced (Comp), Throughput (T), and $dief@t$. Interpretation of all metrics (axes): 'higher is better'.

the versions of ANAPSID produced results continuously compared to MULDER. However, MULDER produces more answers faster than other approaches.

In Fig. 20 reports the performance of approaches using multiple metrics to evaluate the overall performance, completeness and continuous efficiency in time t, i.e., Inverse of Time for the first tuple $(TFFT^{-1})$, Inverse of Total execution time (ET^{-1}), Number of answers produced (Comp), Throughput (T), and $dief@t$, using radar plots. In this plot, the interpretation of the metrics in each axis is **'higher is better'**. As clearly shown, MULDER outperforms ANAPSID-SSGS and ANAPSID-SSGM in almost all metrics and is able to continuously produce results faster. For queries, S1, S6, and S10, all approaches show uniform behavior for all metrics. On the other hand, the performance of both ANAPSID-SSGS and ANAPSID-SSGM on queries, S2, S3, S4, S5, and S7 is almost the same. This is because, LSLOD benchmark datasets do not have sources that share the same RDF-MTs; therefore, the source selection of SSGM is mostly the same as SSGS heuristics. MULDER performs better in all metrics.

We analyze the diefficiency (continuous efficiency) achieved by the two heuristics of ANAPSID and MULDER approach while producing the first k results for LSLOD benchmark queries. Figure 21 reports on the $dief@k$ values while

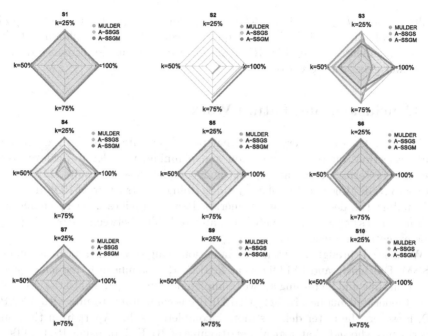

Fig. 21. LSLOD: comparison of diefficiency. Diefficiency while producing a portion of k of answers per LSLOD benchmark query of ANAPSID-SSGS (A-SSGS), ANAPSID-SSGM (A-SSGM), and MULDER query approaches. Performance is measured with $dief@k$, with $k = 25\%, k = 50\%, k = 75\%, k = 100\%$. Interpretation of the axes: 'lower is better'.

producing 25%, 50%, 75%, and 100% of the query results. The value of k is selected by taking the minimum number of results returned from the evaluated approaches. In this plot, the interpretation of the metrics is **'lower is better'**. All approaches perform similar efficiency on the first query, S1. On queries S3, S6, S9, and S10 approaches show different behaviour in different values of k. For instance, on S3 MULDER is able to deliver results faster than ANAPSID-SSGS and ANAPSID-SSGM during the last quarter of overall results, but while k is 25%, 50%, and 75% of the overall results both ANAPSID heuristics delivers faster than MULDER. Even though MULDER performed better on $dief@t$ metrics, as shown in Fig. 20, it produces results slower on k value at 25%, 50%, and 75% on query S3, 50% on S6, 25% on S9, and 75% and 100% of the query results than ANAPSID. On 25%, 75% and 100% of S6 overall query results, 50%, 75%, and 100% of S7 overall query results, 50%, 75%, and 100% of S9 overall query results, and 50% of S10 overall query results all approaches have similar performance of $dief@k$ metrics. For S2, S4, S5, and S7 queries MULDER is able to continuously deliver results faster on all values of k, i.e., 25%, 50%, 75%, and 100% of the overall results and 100% for query S3. In this experiment, we evaluate the continuous efficiency of MULDER compared to ANAPSID. The observed results for both $dief@t$ and $dief@k$ metrics allow us to positively answer **RQ4** and conclude that RDF-MTs based query decomposition and planning techniques implemented in MULDER enable for a continuous performance during the answer generation process.

6   Conclusions and Future Work

MULDER is federated query engine that provides solutions to the problems of source selection, query decomposition and planning, in order to achieve both efficiency in terms of execution time and completeness of results. MULDER resorts to RDF Molecule Templates for describing the structure of RDF datasets and guiding the decomposition of queries. It also provides a query engine for federated access to SPARQL endpoints, able to bridge between parts of a query to be executed in a federated way.

We showed through an extensive evaluation using three different benchmarks (BSBM, FedBench, and LSLOD) that MULDER is significantly reducing query execution time and increasing answer completeness in comparison to the state-of-the-art federated engines. In fact, MULDER has comparable results with ANAPSID, however, the latter delivers more incomplete results. Apart from this, our experiments showed that logical partitioning of RDF data using RDF-MTs is more efficient and effective than using state-of-the-art graph partitioning algorithms, like SemEP and METIS. Furthermore, the analysis of the continuous efficiency of MULDER as an adaptive federated engine demonstrated that it is able to continuously deliver results equal or faster than the ANAPSID adaptive query engine. At the moment, MULDER is able to access only SPARQL endpoints to perform federated queries. In the future, however, we plan to integrate in MULDER additional Web access interfaces, like TPFs (Triple Pattern Fragments) and RESTful APIs. Furthermore, more research can be done in the direction of further optimizing the decomposition and query planning; for instance, to address this, RDF-MTs can be empowered with statistics such as cardinality and link selectivity, in order to reduce, e.g., intermediate results. Finally, we consider RDF Molecule Templates to be an appropriate structure for describing as well as supporting access to other data sources, other than RDF.

Acknowledgements. This work has been partially funded by the EU Horizon 2020 research and innovation programme under the Marie Skłodowska-Curie grant agreement No. 642795 (WDAqua), the EU H2020 programme for the projects BigDataEurope (GA 644564), and iASiS (GA 727658). Mikhail Galkin is supported by a scholarship of German Academic Exchange Service (DAAD).

Appendices

A BSBM Queries

Listing 1.1. Prefixes

```
PREFIX bsbm-inst: <http://www4.wiwiss.fu-berlin.de/bizer/bsbm/v01/instances/>
PREFIX bsbm: <http://www4.wiwiss.fu-berlin.de/bizer/bsbm/v01/vocabulary/>
PREFIX rdfs: <http://www.w3.org/2000/01/rdf-schema#>
PREFIX rdf:  <http://www.w3.org/1999/02/22-rdf-syntax-ns#>
PREFIX rev:  <http://purl.org/stuff/rev#>
PREFIX foaf: <http://xmlns.com/foaf/0.1/>
PREFIX dc:   <http://purl.org/dc/elements/1.1/>
```

Listing 1.2. B1: Find products for a given set of generic features

```
SELECT DISTINCT ?product ?label
WHERE {
        ?product rdfs:label ?label .
        ?product a %ProductType% .
        ?product bsbm:productFeature %ProductFeature1% .
        ?product bsbm:productFeature %ProductFeature2% .
        ?product bsbm:productPropertyNumeric1 ?value1 .
        FILTER (?value1 > %x%)
}
```

Listing 1.3. B2: Retrieve basic information about a specific product for display purposes

```
SELECT ?label ?comment ?producer ?productFeature ?propertyTextual1
?propertyTextual2 ?propertyTextual3
 ?propertyNumeric1 ?propertyNumeric2 ?propertyTextual4 ?propertyTextual5 ?propertyNumeric4
WHERE {
        %ProductXYZ% rdfs:label ?label .
        %ProductXYZ% rdfs:comment ?comment .
        %ProductXYZ% bsbm:producer ?p .
        ?p rdfs:label ?producer .
        %ProductXYZ% dc:publisher ?p .
        %ProductXYZ% bsbm:productFeature ?f .
        ?f rdfs:label ?productFeature .
        %ProductXYZ% bsbm:productPropertyTextual1 ?propertyTextual1 .
        %ProductXYZ% bsbm:productPropertyTextual2 ?propertyTextual2 .
        %ProductXYZ% bsbm:productPropertyTextual3 ?propertyTextual3 .
        %ProductXYZ% bsbm:productPropertyNumeric1 ?propertyNumeric1 .
        %ProductXYZ% bsbm:productPropertyNumeric2 ?propertyNumeric2 .
        OPTIONAL { %ProductXYZ% bsbm:productPropertyTextual4 ?propertyTextual4 }
        OPTIONAL { %ProductXYZ% bsbm:productPropertyTextual5 ?propertyTextual5 }
        OPTIONAL { %ProductXYZ% bsbm:productPropertyNumeric4 ?propertyNumeric4 }
}
```

Listing 1.4. B3: Find products having some specific features and not having one feature

```
SELECT ?product ?label
WHERE {
    ?product rdfs:label ?label .
    ?product a %ProductType% .
    ?product bsbm:productFeature %ProductFeature1% .
    ?product bsbm:productPropertyNumeric1 ?p1 .
    ?product bsbm:productPropertyNumeric3 ?p3 .
    ?product bsbm:productFeature ?pf .
    FILTER (?p3 < %y% )
    FILTER ( ?p1 > %x% )
    FILTER (?pf != bsbm-inst:ProductFeature8001)
}
```

Listing 1.5. B4: Find products matching two different sets of features

```
SELECT DISTINCT ?product ?label ?propertyTextual
WHERE {
    {
    ?product rdfs:label ?label .
    ?product rdf:type %ProductType% .
    ?product bsbm:productFeature %ProductFeature1% .
    ?product bsbm:productFeature %ProductFeature2% .
    ?product bsbm:productPropertyTextual1 ?propertyTextual .
    ?product bsbm:productPropertyNumeric1 ?p1 .
    FILTER ( ?p1 > %x% )
    } UNION {
    ?product rdfs:label ?label .
    ?product rdf:type %ProductType% .
    ?product bsbm:productFeature %ProductFeature1% .
    ?product bsbm:productFeature %ProductFeature3% .
    ?product bsbm:productPropertyTextual1 ?propertyTextual .
    ?product bsbm:productPropertyNumeric2 ?p2 .
    FILTER ( ?p2> %y% )
    }
}
```

Listing 1.6. B5: Find product that are similar to a given product

```
SELECT DISTINCT ?product ?productLabel
WHERE {
    ?product rdfs:label ?productLabel .
    %ProductXYZ% bsbm:productFeature ?prodFeature .
    ?product bsbm:productFeature ?prodFeature .
    %ProductXYZ% bsbm:productPropertyNumeric1 ?origProperty1 .
    ?product bsbm:productPropertyNumeric1 ?simProperty1 .
    %ProductXYZ% bsbm:productPropertyNumeric2 ?origProperty2 .
    ?product bsbm:productPropertyNumeric2 ?simProperty2 .
    FILTER (%ProductXYZ% != ?product)
    FILTER (?simProperty1 < (?origProperty1 + 120) && ?simProperty1 > (?origProperty1 -- 120))
    FILTER (?simProperty2 < (?origProperty2 + 170) && ?simProperty2 > (?origProperty2 -- 170))
}
```

Listing 1.7. B6: Retrieve in-depth information about a specific product including offers and reviews

```
SELECT ?productLabel ?offer ?price ?vendor ?vendorTitle ?review ?revTitle ?reviewer
       ?revName ?rating1 ?rating2
WHERE {
    %ProductXYZ% rdfs:label ?productLabel .
    OPTIONAL {
        ?offer bsbm:product %ProductXYZ% .
        ?offer bsbm:price ?price .
        ?offer bsbm:vendor ?vendor .
        ?vendor rdfs:label ?vendorTitle .
        ?vendor bsbm:country <http://downlode.org/rdf/iso-3166/countries#DE> .
        ?offer dc:publisher ?vendor .
        ?offer bsbm:validTo ?date .
        FILTER (?date > %currentDate% )
    }
    OPTIONAL {
        ?review bsbm:reviewFor %ProductXYZ% .
        ?review rev:reviewer ?reviewer .
        ?reviewer foaf:name ?revName .
        ?review dc:title ?revTitle .
        OPTIONAL { ?review bsbm:rating1 ?rating1 . }
        OPTIONAL { ?review bsbm:rating2 ?rating2 . }
    }
}
```

Listing 1.8. B7: Give me recent reviews in English for a specific product

```
SELECT ?title ?text ?reviewDate ?reviewer ?reviewerName ?rating1 ?rating2 ?rating3 ?rating4
WHERE {
    ?review bsbm:reviewFor %ProductXYZ% .
    ?review dc:title ?title .
    ?review rev:text ?text .
    FILTER langMatches( lang(?text), "EN" )
    ?review bsbm:reviewDate ?reviewDate .
    ?review rev:reviewer ?reviewer .
    ?reviewer foaf:name ?reviewerName .
    OPTIONAL { ?review bsbm:rating1 ?rating1 . }
    OPTIONAL { ?review bsbm:rating2 ?rating2 . }
    OPTIONAL { ?review bsbm:rating3 ?rating3 . }
    OPTIONAL { ?review bsbm:rating4 ?rating4 . }
}
```

Listing 1.9. B8: Get offers for a given product which fulfill specific requirements

```
SELECT DISTINCT ?offer ?price
WHERE {
    ?offer bsbm:product %ProductXYZ% .
    ?offer bsbm:vendor ?vendor .
    ?offer dc:publisher ?vendor .
    ?vendor bsbm:country <http://downlode.org/rdf/iso-3166/countries#US> .
    ?offer bsbm:deliveryDays ?deliveryDays .
    ?offer bsbm:price ?price .
    ?offer bsbm:validTo ?date .
    FILTER (?date > %currentDate% )
    FILTER (?deliveryDays <= 3)
}
```

Listing 1.10. B9: Get all information about an offer.

```
SELECT ?property ?hasValue ?isValueOf WHERE {
{ %OfferXYZ% ?property ?hasValue }
UNION { ?isValueOf ?property %OfferXYZ% }
}
```

Listing 1.11. B10: Get all products' review text and product label from a specific producer

```
SELECT DISTINCT ?product ?revText ?rating3 ?plabel
WHERE {
    ?product rdfs:label ?plabel .
    ?product bsbm:producer %producer1% .
    ?product a ?productType .
    %producer1% a bsbm:Producer .
    %producer1% rdfs:label ?prlabel .
    %producer1% foaf:homepage ?homepage .
    ?review bsbm:reviewFor ?product .
    ?review bsbm:rating1 ?rating1 .
    ?review bsbm:rating2 ?rating2 .
    ?review bsbm:rating3 ?rating3 .
    ?review rev:text ?revText .
}
```

Listing 1.12. B11: Find offer and vendors of a product with a given feature and type

```
SELECT DISTINCT ?product ?producer ?offer ?vendor
WHERE {
    ?product a %ProductType1% .
    ?product rdfs:label ?label .
    ?product bsbm:productFeature %ProductFeature1% .
    ?product rdfs:comment ?productComment .
    ?product bsbm:producer ?producer .
    ?product dc:publisher ?publisher .
    ?product bsbm:productPropertyTextual1 ?propertyTextual1 .
    ?product bsbm:productPropertyTextual2 ?propertyTextual2 .
    ?product bsbm:productPropertyTextual3 ?propertyTextual3 .
    ?product bsbm:productPropertyNumeric1 ?propertyNumeric1 .
    ?product bsbm:productPropertyNumeric2 ?propertyNumeric2 .
    ?producer rdfs:label ?producerLabel .
    ?producer rdfs:comment ?producerComment.
    ?producer dc:publisher ?producerPublisher.
    ?offer bsbm:product ?product .
    ?offer bsbm:price ?price .
    ?offer bsbm:vendor ?vendor .
    ?offer bsbm:validTo %currentDate% .
    ?offer bsbm:validFrom ?offerValidFrom.
    ?offer bsbm:deliveryDays ?offerDeliveryDays.
    ?offer dc:publisher ?offerPublisher.
    ?offer dc:date ?offerPublishDate.
    ?vendor rdfs:label ?vendorLabel.
    ?vendor rdfs:comment ?vendorComment.
    ?vendor bsbm:country ?vcountry.
}
```

Listing 1.13. B12: Find information about its feature, review and reviewer about a given product

```
SELECT DISTINCT ?label ?productType ?productFeature ?producer ?review ?reviewer
WHERE {
    %Product1% a    ?productType .
    %Product1% rdfs:label ?label .
    %Product1% bsbm:productFeature ?productFeature .
    %Product1% rdfs:comment ?productComment .
    %Product1% bsbm:producer ?producer .
    %Product1% dc:publisher ?publisher .
    %Product1% bsbm:productPropertyTextual1 ?propertyTextual1 .
    %Product1% bsbm:productPropertyTextual2 ?propertyTextual2 .
    %Product1% bsbm:productPropertyTextual3 ?propertyTextual3 .
    %Product1% bsbm:productPropertyNumeric1 ?propertyNumeric1 .
    %Product1% bsbm:productPropertyNumeric2 ?propertyNumeric2 .
    ?productFeature rdfs:label ?productFeatureLabel .
    ?productFeature rdfs:comment ?productFeatureComment.
    ?productFeature dc:publisher ?productFeaturePublisher.
    ?review bsbm:reviewFor  %Product1% .
    ?review rev:reviewer ?reviewer .
    ?review dc:title ?revTitle .
    ?reviewer foaf:name ?revName .
    ?reviewer bsbm:country ?country.
    ?reviewer dc:publisher ?reviewerPublisher.
    ?reviewer dc:date ?reviewerPublishDate.
}
```

B LSLOD Queries

Listing 1.14. Prefixes

```
PREFIX drugbank: <http://www4.wiwiss.fu-berlin.de/drugbank/resource/drugbank/>
PREFIX drugcategory: <http://www4.wiwiss.fu-berlin.de/drugbank/resource/drugcategory/>
PREFIX rdf: <http://www.w3.org/1999/02/22-rdf-syntax-ns#>
PREFIX bio2RDF: <http://bio2rdf.org/ns/bio2rdf#>
PREFIX purl: <http://purl.org/dc/elements/1.1/>
PREFIX kegg: <http://bio2rdf.org/ns/kegg#>
PREFIX diseasome: <http://www4.wiwiss.fu-berlin.de/diseasome/resource/diseasome/>
PREFIX dailymed: <http://www4.wiwiss.fu-berlin.de/dailymed/resource/dailymed/>
PREFIX sider: <http://www4.wiwiss.fu-berlin.de/sider/resource/sider/>
PREFIX rdfs: <http://www.w3.org/2000/01/rdf-schema#>
PREFIX owl: <http://www.w3.org/2002/07/owl#>
PREFIX linkedct: <http://data.linkedct.org/resource/>
```

Listing 1.15. S1: Find all drugs along with their indications

```
SELECT ?genericName ?indication
WHERE {
    {
        ?da drugbank:genericName ?genericName.
        ?da drugbank:indication  ?indication.
    }
    UNION {
        ?da dailymed:name ?genericName.
        ?da  dailymed:indication ?indication.
    }
}
```

Listing 1.16. S2: Find all drug descriptions and chemical equations of reactions related to durgs from category Cathartics

```
SELECT ?drugDesc ?cpd ?equation
WHERE {
    ?drug drugbank:drugCategory drugcategory:cathartics.
    ?drug drugbank:keggCompoundId ?cpd .
    ?drug drugbank:description ?drugDesc .
    ?enzyme kegg:xSubstrate ?cpd .
    ?enzyme rdf:type kegg:Enzyme .
    ?reaction kegg:xEnzyme ?enzyme .
    ?reaction kegg:equation ?equation .
}
```

Listing 1.17. S3: Find all drugs, together with the URL of the corresponding Web-pages as well as images

```
SELECT ?drug ?keggUrl ?chebiImage
WHERE {
    ?drug        rdf:type                  drugbank:drugs .
    ?drug        drugbank:keggCompoundId    ?keggDrug .
    ?keggDrug    bio2RDF:url                ?keggUrl .
    ?drug        drugbank:genericName       ?drugBankName .
    ?chebiDrug   purl:title                 ?drugBankName .
    ?chebiDrug   bio2RDF:image              ?chebiImage .
}
```

Listing 1.18. S4: Find KEGG drug names of all drugs in DrugBank belonging to category Micronutrient

```
SELECT distinct ?drug ?title
WHERE {
    ?drug drugbank:drugCategory drugcategory:micronutrient .
    ?drug drugbank:casRegistryNumber ?id .
    ?keggDrug rdf:type kegg:Drug .
    ?keggDrug bio2RDF:xRef ?id .
    ?keggDrug purl:title ?title .
}
```

Listing 1.19. S5: Find all drugs and their mass that affect humans and other mammals. For those having a description of their bioinformation, also return this description

```
SELECT ?drug ?transform ?mass
WHERE {
    ?drug drugbank:affectedOrganism 'Humans_and_other_mammals' .
    ?drug drugbank:casRegistryNumber ?cas.
    ?keggDrug bio2RDF:xRef ?cas .
    ?keggDrug bio2RDF:mass ?mass .
    OPTIONAL { ?drug drugbank:biotransformation ?transform  }
}
```

Listing 1.20. S6: Find diseases and corresponding drugs that target those diseases

```
SELECT ?drug ?disease ?name
WHERE {
    ?drug drugbank:molecularWeightAverage ?weight .
    ?drug drugbank:possibleDiseaseTarget ?disease .
    ?disease diseasome:name ?name .
}
```

Listing 1.21. S7: Find drugs and their side effects with labels for the drug name "Sodium Phosphate" in dailymed

```
SELECT ?drug ?sidereffect ?label
WHERE {
    ?drugAlt sider:sideEffect ?sidereffect .
    ?sidereffect rdfs:label ?label .
    ?drug dailymed:name 'Sodium_Phosphates' .
    ?drug owl:sameAs ?drugAlt .
}
```

Listing 1.22. S8: Find diseases and corresponding drugs that target those diseases along with their labels

```
SELECT ?drug ?disease ?label
WHERE {
    ?disease diseasome:name ?diseasomename .
    ?disease drugbank:possibleDiseaseTarget ?drug .
    ?drug rdfs:label ?label .
}
```

Listing 1.23. S9: Find intervention names and ids for the drugs in dailymed with drug dose, description, inactive ingredients as well as possible disease target

```
SELECT DISTINCT *
WHERE {
    ?intervention a linkedct:intervention.
    ?intervention linkedct:intervention_intervention_name ?intervention_name.
    ?intervention rdfs:seeAlso ?dailymedDrug .
    ?dailymedDrug dailymed:dosage ?dosage.
    ?dailymedDrug  dailymed:description ?description.
    ?dailymedDrug dailymed:inactiveIngredient ?inactiveIngredient .
    ?dailymedDrug dailymed:possibleDiseaseTarget ?possibleDiseaseTarget .
}
```

Listing 1.24. S10: Find intervention names and types for the drugs in durgbank with drug chemical structure, drug state, its protein binding and smiles String Canonical

```
SELECT distinct *
WHERE
{
    ?intervention a linkedCT:intervention.
    ?intervention linkedCT:intervention_name ?intervention_name.
    ?intervention linkedCT:intervention_type ?intervention_type.
    ?intervention rdfs:seeAlso ?drugbankDrug.
    ?drugbankDrug drugbank:structure ?structure.
    ?drugbankDrug drugbank:state ?state.
    ?drugbankDrug drugbank:proteinBinding   ?proteinBinding.
    ?drugbankDrug drugbank:smilesStringCanonical ?smilesStringCanonical .
}
```

References

1. Abdelaziz, I., Essam, M., Mourad, O., Ashraf, A., Kalnis, P.: Lusail: a system for querying linked data at scale. Proc. VLDB Endow. **10**(9), 485–498 (2017)
2. Acosta, M., Vidal, M.-E., Lampo, T., Castillo, J., Ruckhaus, E.: ANAPSID: an adaptive query processing engine for SPARQL endpoints. In: Aroyo, L., Welty, C., Alani, H., Taylor, J., Bernstein, A., Kagal, L., Noy, N., Blomqvist, E. (eds.) ISWC 2011. LNCS, vol. 7031, pp. 18–34. Springer, Heidelberg (2011). https://doi.org/10.1007/978-3-642-25073-6_2

3. Acosta, M., Vidal, M.-E., Sure-Vetter, Y.: Diefficiency metrics: measuring the continuous efficiency of query processing approaches. In: d'Amato, C., et al. (eds.) ISWC 2017, Part II. LNCS, vol. 10588, pp. 3–19. Springer, Cham (2017). https://doi.org/10.1007/978-3-319-68204-4_1

4. Alexander, K., Hausenblas, M.: Describing linked datasets-on the design and usage of VoID, the 'Vocabulary of Interlinked Datasets'. In: LDOW (2009)

5. Basca, C., Bernstein, A.: Querying a messy web of data with Avalanche. J. Web Semant. **26**, 1–28 (2014)

6. Bizer, C., Schultz, A.: The berlin SPARQL benchmark. Int. J. Semant. Web Inf. Syst. (IJSWIS) **5**(2), 1–24 (2009)

7. Charalambidis, A., Troumpoukis, A., Konstantopoulos, S.: SemaGrow: optimizing federated SPARQL queries. In: Proceedings of the 11th International Conference on Semantic Systems, pp. 121–128. ACM (2015)

8. Chen, C., Golshan, B., Halevy, A.Y., Tan, W., Doan, A.: BigGorilla: an open-source ecosystem for data preparation and integration. IEEE Data Eng. Bull. **41**(2), 10–22 (2018)

9. Doan, A., Halevy, A.Y.: Semantic integration research in the database community: a brief survey. AI Mag. **26**(1), 83–94 (2005)

10. Endris, K.M., Galkin, M., Lytra, I., Mami, M.N., Vidal, M.-E., Auer, S.: MULDER: querying the linked data web by bridging RDF molecule templates. In: Benslimane, D., Damiani, E., Grosky, W.I., Hameurlain, A., Sheth, A., Wagner, R.R. (eds.) DEXA 2017. LNCS, vol. 10438, pp. 3–18. Springer, Cham (2017). https://doi.org/10.1007/978-3-319-64468-4_1

11. Feigenbaum, L., Williams, G.T., Clark, K.G., Torres, E.: SPARQL 1.1 protocol. Recommendation, World Wide Web Consortium, March 2013. http://www.w3.org/TR/sparql11-protocol/

12. Fernández, J.D., Llaves, A., Corcho, O.: Efficient RDF interchange (ERI) format for RDF data streams. In: Mika, P., et al. (eds.) ISWC 2014, Part II. LNCS, vol. 8797, pp. 244–259. Springer, Cham (2014). https://doi.org/10.1007/978-3-319-11915-1_16

13. Fernández, J.D., Martínez-Prieto, M.A., de la Fuente Redondo, P., Gutiérrez, C.: Characterising RDF data sets. J. Inf. Sci. **44**(2), 203–229 (2018)

14. Florescu, D., Levy, A.Y., Mendelzon, A.O.: Database techniques for the world-wide web: a survey. SIGMOD Rec. **27**(3), 59–74 (1998)

15. Görlitz, O., Staab, S.: SPLENDID: SPARQL endpoint federation exploiting VoID descriptions. In: COLD (2011)

16. Gubichev, A., Neumann, T.: Exploiting the query structure for efficient join ordering in SPARQL queries. In: EDBT, vol. 14, pp. 439–450 (2014)

17. Halevy, A.Y.: Answering queries using views: a survey. VLDB J. **10**(4), 270–294 (2001)

18. Halevy, A.Y., Rajaraman, A., Ordille, J.J.: Data integration: the teenage years. In: Proceedings of the 32nd International Conference on Very Large Data Bases (VLDB), pp. 9–16 (2006)

19. Hasnain, A., et al.: BioFed: federated query processing over life sciences linked open data. J. Biomed. Semant. **8**(1), 13 (2017)

20. Hayes, P., Patel-Schneider, P.: RDF 1.1 semantics, February 2014

21. Ives, Z.G., Florescu, D., Friedman, M., Levy, A.Y., Weld, D.S.: An adaptive query execution system for data integration. In: SIGMOD 1999, Proceedings ACM SIGMOD International Conference on Management of Data, Philadelphia, Pennsylvania, USA, 1–3 June 1999, pp. 299–310 (1999)

22. Ives, Z.G., Halevy, A.Y., Mork, P., Tatarinov, I.: Piazza: mediation and integration infrastructure for semantic web data. J. Web Sem. **1**(2), 155–175 (2004)
23. Jha, A., et al.: Towards precision medicine: discovering novel gynecological cancer biomarkers and pathways using linked data. J. Biomed. Semant. **8**(1), 40:1–40:16 (2017)
24. Karypis, G., Kumar, V.: A fast and high quality multilevel scheme for partitioning irregular graphs. SIAM J. Sci. Comput. **20**(1), 359–392 (1998)
25. Montoya, G., Skaf-Molli, H., Hose, K.: The *Odyssey* approach for optimizing federated SPARQL queries. In: d'Amato, C., et al. (eds.) ISWC 2017. LNCS, vol. 10587, pp. 471–489. Springer, Cham (2017). https://doi.org/10.1007/978-3-319-68288-4_28
26. Montoya, G., Skaf-Molli, H., Molli, P., Vidal, M.-E.: Federated SPARQL queries processing with replicated fragments. In: Arenas, M., et al. (eds.) ISWC 2015. LNCS, vol. 9366, pp. 36–51. Springer, Cham (2015). https://doi.org/10.1007/978-3-319-25007-6_3
27. Montoya, G., Skaf-Molli, H., Molli, P., Vidal, M.: Decomposing federated queries in presence of replicated fragments. J. Web Semant. **42**, 1–18 (2017)
28. Montoya, G., Vidal, M.-E., Acosta, M.: A heuristic-based approach for planning federated SPARQL queries. In: Proceedings of the Third International Conference on Consuming Linked Data, vol. 905, pp. 63–74. CEUR-WS. org (2012)
29. Neumann, T., Moerkotte, G.: Characteristic sets: accurate cardinality estimation for RDF queries with multiple joins. In: 2011 IEEE 27th International Conference on Data Engineering (ICDE), pp. 984–994. IEEE (2011)
30. Palma, G., Vidal, M.-E., Raschid, L.: Drug-target interaction prediction using semantic similarity and edge partitioning. In: Mika, P., et al. (eds.) ISWC 2014. LNCS, vol. 8796, pp. 131–146. Springer, Cham (2014). https://doi.org/10.1007/978-3-319-11964-9_9
31. Pérez, J., Arenas, M., Gutierrez, C.: Semantics and complexity of SPARQL. ACM Trans. Database Syst. (TODS) **34**(3), 16 (2009)
32. Saleem, M., Khan, Y., Hasnain, A., Ermilov, I., Ngomo, A.N.: A fine-grained evaluation of SPARQL endpoint federation systems. Semant. Web **7**(5), 493–518 (2015)
33. Saleem, M., Ngonga Ngomo, A.-C., Xavier Parreira, J., Deus, H.F., Hauswirth, M.: DAW: Duplicate-AWare federated query processing over the web of data. In: Alani, H., et al. (eds.) ISWC 2013. LNCS, vol. 8218, pp. 574–590. Springer, Heidelberg (2013). https://doi.org/10.1007/978-3-642-41335-3_36
34. Saleem, M., Ngonga Ngomo, A.-C.: HiBISCuS: hypergraph-based source selection for SPARQL endpoint federation. In: Presutti, V., d'Amato, C., Gandon, F., d'Aquin, M., Staab, S., Tordai, A. (eds.) ESWC 2014. LNCS, vol. 8465, pp. 176–191. Springer, Cham (2014). https://doi.org/10.1007/978-3-319-07443-6_13
35. Scheufele, W., Moerkotte, G.: On the complexity of generating optimal plans with cross products. In: 16th ACM SIGACT-SIGMOD-SIGART Symposium on Principles of Database Systems, pp. 238–248 (1997)
36. Schmidt, M., Görlitz, O., Haase, P., Ladwig, G., Schwarte, A., Tran, T.: FedBench: a benchmark suite for federated semantic data query processing. In: Aroyo, L., et al. (eds.) ISWC 2011. LNCS, vol. 7031, pp. 585–600. Springer, Heidelberg (2011). https://doi.org/10.1007/978-3-642-25073-6_37
37. Schmidt, M., Görlitz, O., Haase, P., Ladwig, G., Schwarte, A., Tran, T.: FedBench: a benchmark suite for federated semantic data query processing. In: Aroyo, L., et al. (eds.) ISWC 2011, Part I. LNCS, vol. 7031, pp. 585–600. Springer, Heidelberg (2011). https://doi.org/10.1007/978-3-642-25073-6_37

38. Schmidt, M., Hornung, T., Lausen, G., Pinkel, C.: Sp∧2bench: a SPARQL perfor-
    mance benchmark. In: IEEE 25th International Conference on Data Engineering,
    ICDE 2009, pp. 222–233. IEEE (2009)
39. Schmidt, M., Meier, M., Lausen, G.: Foundations of SPARQL query optimization.
    In: Proceedings of the 13th International Conference on Database Theory, pp. 4–33.
    ACM (2010)
40. Schwarte, A., Haase, P., Hose, K., Schenkel, R., Schmidt, M.: FedX: optimization
    techniques for federated query processing on linked data. In: Aroyo, L., et al. (eds.)
    ISWC 2011. LNCS, vol. 7031, pp. 601–616. Springer, Heidelberg (2011). https://
    doi.org/10.1007/978-3-642-25073-6_38
41. Verborgh, R., et al.: Triple pattern fragments: a low-cost knowledge graph interface
    for the web. J. Web Semant. **37**, 184–206 (2016)
42. Vidal, M., Castillo, S., Acosta, M., Montoya, G., Palma, G.: On the selection of
    SPARQL endpoints to efficiently execute federated SPARQL queries. Trans. Large-
    Scale Data- Knowl.-Centered Syst. **25**, 109–149 (2016)
43. Wylot, M., Cudré-Mauroux, P.: DiploCloud: efficient and scalable management of
    RDF data in the cloud. IEEE Trans. Knowl. Data Eng. **28**(3), 659–674 (2016)
44. Zadorozhny, V., Raschid, L., Vidal, M.-E., Urhan, T., Bright, L.: Efficient evalu-
    ation of queries in a mediator for WebSources. In: Proceedings of the SIGMOD
    Conference, pp. 85–96 (2002)

A Package-to-Group Recommendation Framework

Idir Benouaret[1]([✉]) and Dominique Lenne[2]

[1] Université Jean Monnet Saint Etienne, Laboratoire Hubert Curien UMR 5516,
18 rue Benoit Lauras, 42000 St-Etienne, France
`idir.benouaret@univ-st-etienne.fr`
[2] Sorbonne universités, Université de Technologie de Compiègne,
Heudiasyc – UMR CNRS 7253, Compiègne, France
`dominique.lenne@hds.utc.fr`

Abstract. Recommender systems are important information filtering techniques that retrieve interesting and personalized items for users based on their profiles and past activities. The goal of most recommender systems is to identify a ranked list of items that are likely to be of interest to users. However, there are several applications such as trip planning, where the items to be selected are not intended for single users but for a group of users, and where the group members are interested in package recommendations as collections of items. Recent research on recommender systems has generalized recommendations to suggest packages of items to single users (*Package recommendations*), and single items to groups of users (*Group recommendations*). However, the package-to-group recommendation task has not gained much attention. In this paper, we focus on the task of recommending packages of items to groups of users. This is a task with several real life scenarios, such as recommending a set of Points of Interest packages to tourist groups. We formally define the problem of top-k package-to-group recommendations and propose two models for estimating the preference of a group for a package, incorporating features such as package constraint, user impact and package viability. We design ranking algorithms for finding the top-k package-to-group recommendations and we compare our proposed models with baseline approaches stemming from related works. The experimental evaluation of our proposals, using the Yelp dataset demonstrates that our models find packages of high quality considering important features of package-to-group recommendations.

Keywords: Recommender systems · Package recommendation
Group recommendation · Package-to-group recommendation
Trip planning

1   Introduction

Since the beginning of the web 2.0 era, there has been a massive increase in the volume and availability of data on the Internet. The amount of information in

© Springer-Verlag GmbH Germany, part of Springer Nature 2018
A. Hameurlain et al. (Eds.): TLDKS XXXIX, LNCS 11310, pp. 43–66, 2018.
https://doi.org/10.1007/978-3-662-58415-6_2

the tourism field available in the web has noticed an enormous increase in the last decade. This huge amount of information about tourism and leisure activities has turned the preparation of a trip into a very challenging task. Tourists are starting to have more and more difficulties in finding an interesting set of Points of Interest (POIs) among the huge number of available possibilities in a reasonable amount of time. In the past, people obtained suggestions for their personal trip from their friends or travel agencies. Such traditional sources are user-friendly; however, they have serious limitations. First, the suggestions from friends are limited to the places they have visited before. Second, the information from travel agencies is sometimes biased since agents tend to recommend activities they are associated with. Even worse, when users plan their travel by themselves, they often find their knowledge is too limited to produce a satisfying travel experience.

Search engines partially addressed that problem, however personalization of information was not provided to users. *Recommender systems* (RS) has then emerged as a popular paradigm to solve this problem of information overload. They appeared to be effective in helping users to find what they might be interested in, complementing search engines. Recommender systems are tools that are designed for filtering and sorting items (e.g., movies, books, POIs, etc.). They generally use opinions of a community of users to help individuals in that community to more effectively and efficiently identify items of interest from a potentially overwhelming set of choices [1]. There is a huge diversity of approaches and algorithms one can use to create personalized suggestions. Two of them are very popular: content-based filtering [2] and collaborative filtering [3]. Content-based recommender systems analyze item features and descriptions to identify items that are similar to those the user liked. They perform well when there are sufficient features for items [2]. In the other hand, collaborative filtering recommender systems estimate the similarity between users in order to suggest unseen items to the target user. More often both recommender systems types are combined into so called hybrid recommender systems that helps to reduce the limitation of each method used alone.

However, these *classical* recommender systems are originally confined to recommending *individual* items, e.g., movies, books, etc. to *individual* users. This may not be relevant in some situations. In fact, while traditional research on recommender systems has almost focused on providing recommendations to *single* users, there exist many cases, where the items to be selected are not intended to a personal usage but to a *group* of users. As examples, consider a group of friends seeking to go together on vacation, a family that decide to watch a movie together, or a group of people who want to listen to music in a car. Regardless of how the group is formed, recommending items to the group presents one more challenge in comparison to individual user recommendations [4], which is to identify recommendation that are "good for" the group, i.e., trying to satisfy, as much as possible, the individual preferences of each group member. Existing methods for group recommendations basically follow one of two paradigms [4]. The first, called *profile aggregation*, is to explicitly construct a group profile by combining (aggregating) the profiles of individual members. In this way, the group can be treated as a *virtual* (pseudo) user, and thus standard recommendation techniques can be employed to provide recommendations for the group. The second paradigm, called *recommendation aggregation*, is to first compute recommendations for each member separately, and then employ an aggregation strategy across them to retrieve the group recommendations. Inspired by social choice theory, numerous aggregation strategies for profiles and recommendations exist [5]. This theory studies how individual preferences could be aggregated to reach a collective consensus.

In addition to the group recommendation scenarios, there are several applications where users are interested in packages, i.e., sets (collections) of items, rather than a ranked list of individual items. Examples of such applications may include music [6] (e.g., play list of songs on Last.fm), education [7] (e.g., a set of course combination) and trip planning activities [8,9] (e.g., packages including Points of Interest). Two main challenges rise when dealing with package recommendation. The first one is to define how to "correctly" aggregate items' scores to estimate the score of a whole package. The second challenge is the underlying constraints within the construction of the package. In some applications, the package selection is constrained by the relationships between items in it. For example, in trip planning, the user might specify a budget (e.g., time or cost) and all recommended packages must satisfy these constraints in order to be valid packages. Furthermore, Points of Interest (POIs) cannot be far from each other, otherwise the package would not be satisfactory for the user or even feasible. In a previous paper [10], we designed a framework that recommends to the active user the top-k packages that best correspond to his/her preferences among all valid packages. Our method was inspired from aspects of composite retrieval [11], that group recommendations in packages, where each package is constituted with a set of diverse items and satisfies budget constraint specified by the active user (cost and time).

In the present paper, we merge these two situations (group recommendations and package recommendations). We focus on the task of recommending packages to a group of users. As a motivating example, consider a group of friends who are on vacation, on their first day, they would like to visit a park, to dine at a restaurant and then to have drinks at a nearby bar. Given the potentially overwhelming choices, the group would prefer a recommendation of a (park, restaurant, bar), which is corresponding with the preferences of its members and does not make a group member unhappy with respect to the rest of the group. While there is work on recommending packages of items to a single user (e.g., [6,9,12]) and recommending single items to group of users (e.g., [13,14]) the problem of recommending packages to group of users has not gained much attention. There are studies in helping a group of users to select a package of items (e.g., [15,16]). However, in these works, authors assume that the users are given a set of items and together they decide the items to select. This is more like a group decision making problem where users interact in order to find a consensus, which is a different problem from the package-to-group recommendation problem that we study in this paper. Specifically, given a group of users $\mathcal{G}$, our goal is to suggest one or more packages of items to $\mathcal{G}$, which are suitable for $\mathcal{G}$'s members. This problem has several applications. For example: (i) A summer school with a set of courses that meets the interests of a group of students; (ii) A group of friends who want to package together a set of songs to listen during a trip. Our proposals are not domain-specific, but in this paper, we focus on trip planning, which consists of recommending packages of items (Points of Interest) to a group of users. In line with our previous work on package recommendation [10], we assume the existence of constraints limiting possible item combination that can be included in a package. Constraints that we model can be either hard or soft. We mean by a hard constraint, one which has to be satisfied by a set of items in order to form a valid package (e.g., number of items, cost budget, time budget, etc.). Soft constraints, on the other hand, model desirable but not mandatory properties for a set of items. The constraints we consider in this paper are defined on the basis of relationships between items in a package. In our trip planning scenario, a set of items far from each other is less likely to be chosen as a package by the group of users, than a package of close items. In this case, we say that the package is less *viable*.

Based on the above, we present in this paper two models for computing the package-to-group recommendations. The first model estimates a prediction score that the group of users like individual items (item-to-group), before estimating the prediction score that the group would select a package of items (package-to-group). The second model, first estimates a prediction score for each group member for selecting a specific package (package-to-user), before identifying packages that have high probability to be selected by the group (package-to-group). In addition, we design and implement efficient ranking algorithms for our models to retrieve the top-k package-to-group recommendations using the *Produce-and-choose* paradigm.

The roadmap of the paper is as follows. In Sect. 2, we give a formalization of the package-to-group recommendation problem that we study. In Sect. 3, we introduce some preliminaries of our work. Section 4 presents our two models for estimating the package-to-group score and introduces package viability. In Sect. 5, we describe our algorithms that implement our models for retrieving the top-k package-to-group recommendations. In Sect. 6, we subject our proposals to experimental analysis using a real dataset and investigate the quality of the recommended packages. In Sect. 7 we review related works on package recommendations and group recommendations. Finally, Sect. 8 concludes the paper and presents future work.

2   Problem Statement

We assume that we are given a collection of items $\mathcal{I}$ and a collection of users $\mathcal{U}$. Each user express his/her preferences to items from $\mathcal{I}$ through ratings. We denote by $r(u, i)$ the rating of user u for item i, this rating may be *explicit*, i.e., u has explicitly consumed and rated item i, or *implicit*, i.e., the rating is predicted using a classical recommendation algorithm (e.g., collaborative filtering [17]).

Items from which packages are computed belong to categories. For example, POIs can be classified to museums, landmarks, parks, etc. Thus, we assume that we have a set of categories $\mathcal{C}$, $|\mathcal{C}| > 1$, (e.g., $\mathcal{C} = \{museum, landscape, restaurant\}$) and that each item $i \in \mathcal{I}$ belongs to one or more categories in $\mathcal{C}$. We note C_i, $C_i \subset \mathcal{C}$, the set of all categories of item i.

Given a *group* (set) of users $\mathcal{G} \subset \mathcal{U}$, our goal is recommend to $\mathcal{G}$ the top-k packages $\{P_1, P_2, ..., P_k\}$ that best satisfy the group members. Each package P_i is a set of items in $\mathcal{I}$. We assume that each item can appear at most once in the recommended packages, i.e., $P_1 \cap P_2 \cap ... \cap P_k = \emptyset$. Recommended packages must be *valid*. i.e., have specific properties.

Definition 1 (Valid package). *Given a set-valued non negative and monotone function $f : 2^{\mathcal{I}} \rightarrow \mathbb{R}_+$, and given a budget threshold β specified by the group, a package $P \subset \mathcal{I}$ is said to be valid iff $f(P) \leq \beta$.*

Our proposals do not depend on a specific budget constraint. The f function is generic, which allows us to take into account different aspects of the budget constraint. Typical examples of budget are simply the number of items forming a package, or an upper-bound on the sum of the costs of items forming a package, while f is the sum of the average price induced by visiting items in the package. Another example of budget could be the available time for visiting all items in a package, while f is the sum of the average visiting time of each item in the package.

Formally, the top-k package-to-group recommendation task takes as input a group of users $\mathcal{G} \subset \mathcal{U}$, a set of items $\mathcal{I}$, a set of ratings for each group member, a group specified budget β, and has to recommend to $\mathcal{G}$ the top-k packages $\{P_1, P_2, ..., P_k\}$ that are *Valid*, i.e., $\forall i, f(P_i) \leq \beta$ and that are most preferable among all *Valid* packages.

3   Preliminaries

3.1   Category Distance and Similarity Between Items

Our distance between items is based on a taxonomy of hierarchical topic categories organized in a tree structure. Formally, we used a domain ontology developed by [18] to represent these categories. Figure 1 shows a part of this taxonomy.

Let $\mathcal{I}$ be the set of all possible items for potential package recommendations. Each item $i \in \mathcal{I}$ is associated to one or more categories in the taxonomy, e.g., $\{restaurant, bar, nightlife\}$ We define the category distance $dist_c : \mathcal{I} \times \mathcal{I} \rightarrow \mathbb{N}$ between two items i and j as the length of the shortest path between the two closest categories of i and j in the taxonomy. Formally, we denote by C_i and C_j the set of all categories of item i, respectively j. The category distance between items i and j is calculated using the following formula:

$$dist_c(i,j) = min_{c_i \in C_i, c_j \in C_j} dist_{edge}(c_i, c_j) \tag{1}$$

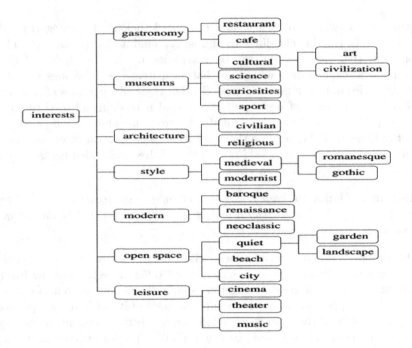

Fig. 1. Part of the taxonomy [18] used to represent item categories.

where $dist_{edge}$ is the function computing the shortest path (number of edges) between two nodes in the taxonomy.

The similarity between two items i and j can then be estimated as a function of their category distance. This similarity evaluates to what extent two items i and j deal with the same thematic. The greater the category distance between two items is, the smaller the similarity between these two items is and vice versa. For example, an item which is a restaurant and a bar is similar to an item which is a cafe. Formally, the similarity between two items i and j is calculated as follows:

$$sim(i,j) = \frac{1}{1 + dist_c(i,j)} \qquad (2)$$

3.2 Individual Recommendations

Let $\mathcal{U}$ denote the set of users and $\mathcal{I}$ denote the set of items (e.g., movies, travel destinations, restaurants) in the system. Each user u may have provided a rating for an item i in the range of 0 to 5, which is denoted as $rating(u,i)$. We further generate relevance scores for each pair of user and item, denoted as $r(u,i)$. This relevance score comes from two sources. If the user has provided a rating for the item, then it is simply the user provided rating. Otherwise, the system generates the relevance score using a recommendation strategy.

One of the most well-known recommendation strategies relies on finding items similar to the users previously highly rated items (item-based collaborative filtering). The relevance of an item $i \in \mathcal{I}$ by an active user $u \in \mathcal{U}$ is commonly estimated as follow:

$$r(u, i) = \sum_{j \in \mathcal{I}_u} sim(i, j) \times rating(u, j) \tag{3}$$

where $\mathcal{I}_u$ is the set of items rated by user u and $sim(i, j)$ is the similarity between items i and j.

4   Models

In this section we present our two models for the package-to-group recommendation task. Given a target group of users $\mathcal{G}$ with the preference background of every group member, the objective is to estimate a prediction score that we denote by $Score(\mathcal{G}, \mathcal{P})$. This score measures the probability that the group $\mathcal{G}$ will like and select the package $\mathcal{P}$. This score obviously depends on the preference of each user $u \in \mathcal{G}$ for the individual items $i \in \mathcal{I}$. Given a user $u \in \mathcal{G}$ and an item $i \in \mathcal{I}$, we denote $Score(u, i)$ the prediction score which estimates to what extent the user u might like item i and select it. In other words, we need to calculate the probability that the user u will choose item i rather than other items from the set of items $\mathcal{I} \setminus \mathcal{I}_u$ that the user has not rated yet, where $\mathcal{I}_u$ is the set of items already rated explicitly by the user u. This score can be defined as:

$$Score(u, i) = \frac{r(u, i)}{\sum_{j \in \mathcal{I} \setminus \mathcal{I}_u} r(u, j)} \tag{4}$$

In formula 4, $r(u, i)$ is the rating (explicit or implicit) of user u for item i. Intuitively, user u is more likely to accept a recommendation of an item i with higher predicted rating $r(u, i)$ compared to $r(u, j)$ for other items $j \in \mathcal{I} \setminus \mathcal{I}_u$.

Next, we present two models for computing $Score(\mathcal{G}, \mathcal{P})$ based on the individual scores $Score(u, i)$ and other factors, such as the influence between users in the group.

4.1   Item to Group-Package to Group (IG-PG) Model

The IG-PG model works with two steps. We first define the score that estimates the probability that the group $\mathcal{G}$ will select an item i. Then we combine the scores of individual items to estimate the score of the package according to the group $\mathcal{G}$ and items in the target package $\mathcal{P}$.

Item-to-Group Score: Given a group $\mathcal{G}$ and a target item $i \in \mathcal{I}$, we have to define $Score(\mathcal{G}, i)$, the item-to-group score, which estimates the probability of group $\mathcal{G}$ selecting item i, that reflects the interests and preferences of all group members on a single item i. In general, group members may not always have the same tastes and a consensus score for each item needs to be carefully designed. This score can be computed solely by aggregating individual scores of each group member using a well known aggregation function, such as *average* or *least-misery* aggregations. However, these strategies have generally some drawbacks. First, they assume that the aggregation strategy is fixed and is the same for all items. They often treat users in the group equally and the same as individual users, mostly assuming that the behavior and preferences of users in groups is identical to the ones of individual users. However, recent research on group recommender systems demonstrates that better results were obtained when modeling users interactions within the group [19,20]. In our model, we adopt this approach where it is assumed that different group members have different impact on the group's decision making process. In other words, one or more members of the group, who could be considered as experts on a specific category, may influence the group in selecting an item in this category. For example, the preference of a group member who is a "museum lover" will count more than other users in selecting a specific museum.

Following this intuition, we model the group preference with a weighted average strategy, depending on the target item. In other words, we average individual user scores for the target item i, weighted by the impacts of users in the group. Each user $u \in \mathcal{G}$ has an importance factor $w_{u,i}$ estimating the impact of the user u in selecting item i. Therefore, we have:

$$Score(\mathcal{G}, i) = \sum_{u \in \mathcal{G}} w_{u,i} \times Score(u, i) \tag{5}$$

In this work, we assume that the user impact $w_{u,i}$ in selecting item i is proportional to the activity of that user in the set of categories C_i of item i, relatively to the other group members in the group. This captures the relative *expertise* of the user in the group for items in these categories, which determines his/her influence in the group when selecting this item. Specifically, following this assumption, we estimate $w_{u,i}$ using the number of explicit ratings user u has given for items that are similar to the target item i. Specifically, let $\lambda_{u,i}$ denote the number of explicit ratings user $u \in \mathcal{G}$ has given for items that are similar to item i. We assume that item i is similar to item j if their similarity is greater than a threshold, i.e., $sim(i, j) > \varepsilon$. We consider that:

$$w_{u,i} = \frac{\lambda_{u,i}}{\sum_{v \in \mathcal{G}} \lambda_{v,i}} \tag{6}$$

We note that Eq. 5 is general enough to model different scenarios, depending on the definition we want to give for the weight $w_{u,i}$. For example, we can set the weights to the uniform distribution over group members, so that all users will has the same impact and will influence equally in the selection process.

Package-to-Group Score: Once we know how to estimate for each item i its item-to-group score: $Score(\mathcal{G}, i)$. We can derive the package-to-group score, which estimates the probability that group $\mathcal{G}$ will select package $\mathcal{P}$. If we assume that each item i in the package $\mathcal{P}$ is chosen independently from others items j in the same package, therefore, given $Score(\mathcal{G}, i)$ we have:

$$Score(\mathcal{G}, \mathcal{P}) = \prod_{i \in \mathcal{P}} Score(\mathcal{G}, i)$$
$$= \prod_{i \in \mathcal{P}} \left(\sum_{u \in \mathcal{G}} w_{u,i} \times Score(u, i) \right) \tag{7}$$

4.2 Package to User-Package to Group (PU-PG) Model

In the IG-PG model, we assumed that items in the package were selected on the basis of the group preference over individual items. This selection was made according to the preference of each group member weighted by the user impact in selecting a specific item. Once the group preference was estimated for items we derived the package-to-group score.

In the PU-PG model, this generative process is reversed. We first estimate the probability score that each group member $u \in \mathcal{G}$ will select a target package $\mathcal{P}$, then according to each group member score, we derive a whole package-to-group score.

Package-to-User Score. Here, the score of the package depends only on a specific user u according to his own preferences. Obviously the prediction score of a package depends on the prediction score of individual items forming this package. If we assume that each item i in the package $\mathcal{P}$ is chosen independently of other items in the same package, therefore, given $Score(u, i)$ we have:

$$Score(u, \mathcal{P}) = \prod_{i \in \mathcal{P}} Score(u, i) \tag{8}$$

Package-to-Group Score. Once we have estimated the score probability that each group member $u \in \mathcal{G}$ will select a target package $\mathcal{P}$, we estimate the global package-to-group score according to each group member. Similar to the IG-PG model, we assume that each user in the group will have a different user impact during the selection process, according to his relative "expertise". What is different in the PU-PG model is that we consider the user impact on packages instead of items. Thus, we model the package-to-group score as a weighted average strategy, where every package-to-user score is weighted by the user impact in selecting this package. Formally, we have:

$$Score(\mathcal{G}, \mathcal{P}) = \sum_{u \in \mathcal{G}} w_{u,\mathcal{P}} \times Score(u, \mathcal{P})$$
$$= \sum_{u \in \mathcal{G}} w_{u,\mathcal{P}} \times \left(\prod_{i \in \mathcal{P}} Score(u, i) \right) \tag{9}$$

In this model, the user impact in the decision process is based on the influence of each user u on all target items in the package $\mathcal{P}$ collectively. Therefore, we adapt weights in Eq. 6 to take into account all items in the package.

$$w_{u,\mathcal{P}} = \frac{\sum_{i \in \mathcal{P}} \lambda_{u,i}}{\sum_{v \in \mathcal{G}} \sum_{i \in \mathcal{P}} \lambda_{v,i}} \tag{10}$$

In this model, the user impact in selecting a package depends on the number of explicit ratings that he gave for similar items to those forming the package.

To sum up, the PU-PG model computes the package-to-group score in a two-step process. First, the package-to-user score computes the probability $Score(u, \mathcal{P})$ that each user $u \in \mathcal{G}$ selects package $\mathcal{P}$. Second, the package-to-group phase computes the overall group preference $Score(\mathcal{G}, \mathcal{P})$ depending on the preference of each user weighted by the user impact weights.

Note that the PU-PG model gives more power to the user with a high user impact, since the package selection is according to each group members. As a result, IG-PG and PU-PG may produce very different package selection scores, even in the case of uniform user impact weights. Let's consider an example (Table 1), where a group of two users $\mathcal{G} = \{u_1, u_2\}$ wants to select a package of two items ($|\mathcal{P}| = 2$), over four candidate items, $\mathcal{I} = \{i_1, i_2, i_3, i_4\}$, with the individual item scores shown in the Table 1(a). Let's assume that the user impacts are uniform, i.e., $w_{u_1,i} = w_{u_2,i} = 1/2$ for every item i and $w_{u_1,\mathcal{P}} = w_{u_2,\mathcal{P}} = 1/2$ for every possible package $\mathcal{P}$. Table 1b shows the computed scores of IG-PG and PU-PG models for four possible packages.

Table 1. Comparison between IG-PG and PU-PG models

items/users	u_1	u_2
i_1	1	0
i_2	0	1
i_3	1	0
i_4	0	1

(a) $Score(u, i)$

Packages	IG-PG	PU-PG
$P_1 = \{i_1, i_3\}$	1/4	1/2
$P_2 = \{i_1, i_4\}$	1/4	0
$P_3 = \{i_2, i_3\}$	1/4	0
$P_4 = \{i_2, i_4\}$	1/4	1/2

(b) $Score(\mathcal{G}, \mathcal{P})$

As we pointed it out before, the example shows that IG-PG and PU-PG models may produce very different packages. In the example, for the PU-PG model, a package that no user likes as a whole (P_2 and P_3) will have a very small (zero, in the example) score, while the packages with high score are favored by individual preferences (P_1 is favored by u_1 and P_4 is favored by u_2). On the other hand, the IG-PG model produces a reasonable score as long as there are some users that like some items in the package (e.g., u_1 likes i_1 and u_2 likes i_4), the IG-PG model is more fair in balancing the preferences of each group member. As a consequence, in the example (Table 1), the PU-PG model favors packages $P_1 = \{i_1, i_3\}$ (u_2 dislikes i_1 and i_3) and $P_4 = \{i_2, i_4\}$ (u_1 dislikes i_2 and i_4), when the IG-PG model produces the same score for the four packages P_1, P_2, P_3 and P_4.

4.3 Package Viability

From the beginning, we have considered only the preferences of each group member over the items in a target package when estimating the group to package score (Eqs. (7) and (8)). So far, we assumed that there is no dependence between items and that items in the package are selected independently. However, in our trip planning scenario, this assumption may not be true. In fact, in a real life scenario with real tourists, some items are more likely to be chosen together as a whole package than other items. For example, let's assume that a group of tourists in a city wants to first visit a museum, then to have dinner at a restaurant and finally to have drinks at a bar. A package with a museum, a restaurant and a bar has higher chances to be preferred by the group if these items are geographically close. Instead, the package will be less viable if the items forming the package are situated in opposite cites of the city. Motivated by this assumption, we define a measure of viability $\mathcal{V}(\mathcal{P})$ which estimates to what extent a package $\mathcal{P}$ is viable as a whole, assuming that items are not independent of each other. Following this intuition, we want to have a measure of viability $\mathcal{V}(\mathcal{P})$ which is inversely proportional to the average pairwise distance between all items in the package $\mathcal{P}$.

Assuming that we have a function $dist(i,j)$ which calculates the geographical distance between two items i and j, the average pairwise distance of a package $\mathcal{P}$ is defined by:

$$AVG_{dist}(\mathcal{P}) = \frac{2}{|\mathcal{P}|.(|\mathcal{P}|-1)} \sum_{i,j \in \mathcal{P}, i \neq j} dist(i,j) \qquad (11)$$

Hence, we can define the viability of a package as follows:

$$\mathcal{V}(\mathcal{P}) \propto e^{-AVG_{dist}(\mathcal{P})} \qquad (12)$$

Intuitively, if the average pairwise distance between all pairs of items in a package is large, the package has a low probability to be viable and to be selected by the group.

Now that we have defined the viability of a package, we can formalize the probability score that group $\mathcal{G}$ will select a package $\mathcal{P}$ and that the selected package is viable.

$$Score_v(\mathcal{G}, \mathcal{P}) = \mathcal{V}(\mathcal{P}) \times Score(\mathcal{G}, \mathcal{P}) \qquad (13)$$

We note that both our models IG-PG and PU-PG are augmented with the package viability score that we just defined.

5   Recommending Top-K Packages

Given a group of users $\mathcal{G}$, a set of items $\mathcal{I}$ and each user ratings over the items, our goal is to recommend the top-k packages with the k highest $Score_v(\mathcal{G}, \mathcal{P})$ according to Eq. (13) for both IG-PG and PU-PG models. Note that the number of candidate packages is exponential to the number of available items $|\mathcal{I}|$. So, it is obvious that computing the scores of every possible candidate packages cannot be done in a polynomial time. Thus, we need an efficient way to retrieve the top-k packages. In line with our previous work [10], we tackle the problem of package recommendations by generating a set of "good" candidate packages and then selecting the best possible subset. This approach uses the paradigm *Produce-and-choose* introduced by [11], which we choose for solving our problem. The construction of top-k packages for either the IG-PG model or the PU-PG model is done in two phases: first, a set of valid packages are produced in large quantities with a cardinality $c \gg k$, packages are formed by aggregation around a pivot item. After that, we pick the top-k packages that achieved the k highest scores. Our approach for forming a set of good valid packages is inspired from the composite retrieval algorithm "BOBO" (*Bundles One-By-One*) introduced by Amer-Yahia *et al* [11], which itself is inspired from $k - nn$ clustering. We adopted the idea of this algorithm to take into account our models and scoring functions.

5.1 Algorithms for the IG-PG Model

Recall that the IG-PG model includes two steps: the item-to-group step which estimates the score of each item according to the group preferences, and the package-to-group step which combines items into packages. The final scoring function used by IG-PG model (Eq. (13)) considers two features in the package-to-group step: the group preference score (Eq. (7)) and the package viability (Eq. (12)). This means that combining the best items found in the item-to-group step into the recommended packages will not necessarily lead to the top-k packages.

We propose a 2-level algorithm BOBO-IG-PG, which implements the item-to-group and package-to-group steps sequentially. First, we calculate the item-to-group $Score(\mathcal{G}, i)$ according to the group preferences, for each candidate item i. Then, the package-to-group step takes as input a ranked list of the candidate items according to their respective item-to-group score and combine these items into candidate packages. Once we have a sufficient number of good and valid packages, we select the k packages with the highest scores.

More precisely, our algorithm takes as input a group of user $\mathcal{G}$, a set of items $\mathcal{I}$, a budget specified by the group β. Our algorithm starts by removing from the candidate items $\mathcal{I}$ all the items that are already rated by each group member $u \in \mathcal{G}$ (line 1 to line 3). After that, we calculate the item-to-group score for each candidate item $i \in \mathcal{I}$ according to Eq. 5 (line 4 to line 6). After this initialization step of the algorithm, we start with an empty set of packages (line 7). Then, a list of candidates item pivots is constructed (line 8), this list of potential items that will include the packages is ranked in a decreasing order with respect to the item-to-group score. Then, as long as the number of formed packages are less than the number of required packages c and the *Pivots* list is not empty, at each iteration, we pick the first item from the set of *Pivots* (line 10), and a package is built around it (line 12). This is done by the routine *Pick_Group_Package* described

---

Algorithm 1. BOBO-IG-PG

Input: a group $\mathcal{G}$, a set of items $\mathcal{I}$, a budget β, number of packages c
Output: Top-k packages
1 **foreach** $u \in \mathcal{G}$ **do**
2 | $\mathcal{I} \leftarrow \mathcal{I} \setminus I_u$
3 **end**
4 **foreach** *item* $i \in \mathcal{I}$ **do**
5 | calculate $Score(\mathcal{G}, i)$ // Eq. (5)
6 **end**
7 $Packages \leftarrow \emptyset$
8 $Pivots \leftarrow Descending\_sort(\mathcal{I}, Score(\mathcal{G}, i))$
9 **while** $(Pivots \neq \emptyset)$ **and** $|Packages| < c$ **do**
10 | $w \leftarrow Pivots[0]$
11 | $Pivots \leftarrow Pivots - \{w\}$
12 | $P \leftarrow Pick\_Group\_Package(w, \mathcal{G}, \mathcal{I}, \beta)$
13 | $Packages \leftarrow Packages \cup P$
14 | $Pivots \leftarrow Pivots - P$
15 **end**
16 **return** *Top-k Packages*

---

in Algorithm 2. This routine takes as input the current set of candidate items $\mathcal{I}$, the group budget β and the pivot item selected w, and outputs a valid package P. First, we add to the package the pivot item w (line 1) and we remove it from candidate items $\mathcal{I}$ (line 2). The routine works in a greedy fashion, it keeps picking the next item from the set of active items that will maximize the score of the package (according to Eq. (13) which is being created around the selected pivot (line 4), as far as the budget constraint is satisfied (line 5). If the next selected item respects the budget constraint then it is added to the current package (line 6). It is then discarded from the *active* items (line 11), so that it will not appear in any other package. Note that, without loss of generality, we assume that for all items $i \in \mathcal{I}$, $f(i) < \beta$. Let's go back to BOBO-IG-PG's main loop: once a candidate package is created around the selected pivot, it is added to "*Packages*" (line 13) and its elements are removed from "*Pivots*" (line 14) so that they are not longer used.

Once the required number of packages has been created, they are ranked following their respective scores. Afterwards, the k packages having the best scores are selected as the top-k packages.

5.2 Algorithms for the PU-PG Model

We can design algorithms for the PU-PG model using a similar manner as for the IG-PG model. Its adaptation is straightforward. Recall that the PU-PG model includes two steps: the package-to-user step which estimates the score of a package to be selected by an individual user ($Score(u, \mathcal{P})$), and the package-to-group step which estimates the score that a package will be selected by the

---

Algorithm 2. PICK_GROUP_PACKAGE

Input: pivot item w, group $\mathcal{G}$, candidate items $\mathcal{I}, budget\beta$
Output: a valid package $\mathcal{P}$

1 $\mathcal{P} \leftarrow w$
2 $active \leftarrow \mathcal{I} - \{w\}$
3 **while** *(not finish)* **do**
4 | $i \leftarrow argmax_{i \in active} Score_v(\mathcal{G}, \mathcal{P} \cup \{i\})$
5 | **if** $(f(\mathcal{P} \cup \{i\}) \leq \beta)$ **then**
6 | | $\mathcal{P} \leftarrow \mathcal{P} \cup \{i\}$
7 | **end**
8 | **else**
9 | | $finish \leftarrow true$
10 | **end**
11 | $active \leftarrow active - i$
12 **end**
13 **return** $\mathcal{P}$

---

group $(Score(\mathcal{G}, \mathcal{P}))$, according to the individual preferences of each user u for this package.

Similar to the IG-PG model we also propose a 2-level algorithm for the PU-PG model, which implements the package-to-user and package-to-group sequentially. In a first step, we find for each user $u \in \mathcal{G}$ the relevant packages $\mathcal{P}_u$ and their scores according to his own preferences, i.e., $Score(u, \mathcal{P})$ in Eq. (8). Then we select the top-k packages from the set of all recommended packages for each user $u \in \mathcal{G}$, i.e., $\cup_{u \in \mathcal{G}} \mathcal{P}_u$.

6 Experiments

In this section, we subject our package-to-group models and algorithms to an experimental evaluation and measure the effectiveness of the proposed solutions.

6.1 Dataset

We use the Yelp challenge dataset[1] in our evaluation. The items in this dataset correspond to points of interest in 11 metropolitan areas in USA, rated by Yelp users. The original Yelp dataset contains about $100K$ users, $174K$ items and $5.2M$ reviews (with a numerical rating between 1–5 stars). In order to make our setup more realistic, we choose items from a single city: "Phoenix". Note that, in our setup we excluded items that received less than 20 ratings, and users that have given less than 20 explicit ratings. Even though, the rating matrix is still very spare. To overcome this problem, we employ collaborative filtering (CF) [17] to get additional predicted ratings for each user in order to fill the rating

---

[1] https://www.yelp.com/dataset/challenge.

Algorithm 3. BOBO-PU-PG

Input: a group $\mathcal{G}$, a set of items $\mathcal{I}$, a Budget β, number of packages c
Output: Top-k packages
1 **foreach** $u \in \mathcal{G}$ **do**
2 | $\mathcal{I} \leftarrow \mathcal{I} \setminus I_u$
3 **end**
4 $Packages_\mathcal{G} \leftarrow \emptyset$
5 **foreach** $u \in \mathcal{G}$ **do**
6 | $\mathcal{P}_u \leftarrow \emptyset$
7 | $Pivots_u \leftarrow Descending\_sort(\mathcal{I}, Score(u, i))$
8 | **while** $(Pivots_u \neq \emptyset)$ **and** $|\mathcal{P}_u| < c$ **do**
9 | | $w \leftarrow Pivots_u[0]$
10 | | $Pivots_u \leftarrow Pivots_u - \{w\}$
11 | | $\mathcal{P} \leftarrow$ Pick_User_Package$(w, u, \mathcal{I}, \beta)$
12 | | $\mathcal{P}_u \leftarrow \mathcal{P}_u \cup \mathcal{P}$
13 | | $Pivots_u \leftarrow Pivots_u - \mathcal{P}$
14 | **end**
15 | $Packages_\mathcal{G} \leftarrow Packages_\mathcal{G} \cup \mathcal{P}_u$
16 **end**
17 **foreach** $\mathcal{P} \in Packages_\mathcal{G}$ **do**
18 | Calculate $Score_v(\mathcal{G}, \mathcal{P})$
19 **end**
20 **return** *Top-k Packages*

Algorithm 4. PICK_USER_PACKAGE

Input: pivot item w, user u, candidate items $\mathcal{I}$, budget β
Output: a valid package $\mathcal{P}$
1 $\mathcal{P} \leftarrow w$
2 $active \leftarrow \mathcal{I} - \{w\}$
3 **while** $(not\ finish)$ **do**
4 | $i \leftarrow argmax_{i \in active} Score(u, \mathcal{P} \cup \{i\})$ // Eq. (8)
5 | **if** $(f(\mathcal{P}_u \cup \{i\}) \leq \beta)$ **then**
6 | | $\mathcal{P} \leftarrow \mathcal{P} \cup \{i\}$
7 | **end**
8 | **else**
9 | | $finish \leftarrow true$
10 | **end**
11 | $active \leftarrow active - i$
12 **end**
13 **return** $\mathcal{P}$

---

matrix as much as possible. In particular, we use Recommendelab[2] to build an item-based collaborative filtering and retrieve for every user u item ratings that are not present in the dataset. For the items that are neither explicitly rated by

---

[2] https://cran.r-project.org/web/packages/recommenderlab/.

u in the dataset nor recommended using collaborative filtering, we set zero as u's rating. Finally, we end up having a rating matrix with about $3.2M$ non-zero ratings in total, 2348 users and 1403 items.

6.2   Group Generation

In order to conduct the experiments, we need to generate artificial groups from our dataset. The key factor we consider is group cohesiveness (similarity). Groups of varying sizes and similarities were generated by sampling our dataset. We considered groups containing one, two, four, six or eight users. Moreover, we distinguished between three categories of groups to conduct the experiments: *random user groups* (RAND Groups), *Similar user groups* (SIM Groups) and *Diverse user groups* (DIV Groups). These are cases that are common in real life scenarios. Groups with highly similar members represent people with common tastes, such as groups of friends sharing the same interests. Groups with highly dissimilar members represent people with opposite tastes. Whereas, random groups represent people without any social relations, such as random people going to the same summer camp.

Similarity between users is computed using the Pearson Correlation coefficient (PCC). We defined two group similarity levels: 0.3 and 0.7. Groups with high inner group similarity are defined as those containing users with user-to-user similarity higher than 0.7. For example to form a group of three similar users, we select three users from the dataset u_1, u_2, u_3, such that $\forall i, j, sim(u_i, u_j) > 0.7$, where $1 \leq i, j \leq 3, i \neq j$. On the other hand, diverse user groups are defined as those containing users with user-to-user similarity less than 0.3. Random groups were formed without considering any restriction on the user-to-user similarity. When forming the groups, in order to measure if two users were similar or not we considered only pairs of users that have rated at least 5 common items. This is a common practice in memory based collaborative filtering literature and assures that the computed similarity value is reliable and the correlation is not high or low just by chance, i.e., because there is a perfect agreement or disagreement on a small set of items. In our experiments we report average values over 20 different group initializations.

6.3   Experimental Protocol and Quality Metrics

Our goal is to test the effectiveness of our models according to the quality of the recommended packages. To this end, we compared several versions of our models, corresponding to different possible combinations of our parameters. Table 2 summarizes all parameters involved in our experiments. On each running test, we vary one parameter while keeping the others to their default values. Each running test computes the top-k recommended packages to a group of users $\mathcal{G}$. We note that Yelp dataset does not provide neither items cost nor items average visiting time. Thus, we consider the group specified budget constraint to be simply the number of items forming a package $|\mathcal{P}|$.

Table 2. Experimental parameters (default values in bold)

Description	Parameter	Values
Group size	$|\mathcal{G}|$	1,2,**4**,6,8
Package size	$|\mathcal{P}|$	1,3,**5**,7,10
Number of recommended packages	k	1,5,**10**,15,20
Group similarity	Type (G)	SIM, **RAND**, DIV

We now study the effectiveness of our proposed IG-PG and PU-PG models according to the quality of the recommended packages as well as their viability. In the evaluation, we include two baselines approaches which are based of the state-of-the-art group recommendation techniques: Average aggregation and Least misery aggregation. Average and Least-Misery aggregation models are considered because they are the most prevalent mechanisms being employed [13]. In the Average strategy, the group score of an item i is calculated as the average of the individual ratings of each group member, i.e., $Score(\mathcal{G}, u) = \frac{\sum_{u \in \mathcal{G}} r(u,i)}{|\mathcal{G}|}$. In the Least-Misery strategy, the group score for an item i is equal to the smallest rating for i in the group, i.e., $Score(\mathcal{G}, u) = Min_{u \in \mathcal{G}} r(u, i)$. Thus, each item is predicted to be liked by the group as it is by the less satisfied member. Least-Misery model captures cases where some user has a strong preference (e.g., a vegetarian who cannot go to a steakhouse) and that user's preference acts as a veto. These two models where originally designed for group recommendations involving individual items, so we adapted these models into our package-to-group problem. More specifically, we study the two following baseline algorithms:

$AVG - GREEDY$: A greedy algorithm that selects items greedily to maximize the average rating of the package

$$AVG(\mathcal{G}, \mathcal{P}) = \frac{1}{|\mathcal{G}||\mathcal{P}|} \cdot \sum_{u \in \mathcal{G}} \sum_{i \in \mathcal{P}} r(u, i)$$

$LM - GREEDY$: A greedy algorithm that selects items greedily to maximize the least misery value of the package

$$LM(\mathcal{G}, \mathcal{P}) = Min_{u \in \mathcal{G}} Min_{i \in \mathcal{P}} r(u, i)$$

We also experimented with a random selection of items. This algorithm is consistently outperformed by all other algorithms, so we do not include it in the experiments to avoid overloading the plots.

We compare IG-PG, PU-PG, AVG-GREEDY, LM-GREEDY in terms of package quality using two metrics: the first metric is the average group-item rating $R(\mathcal{G}, \mathcal{P})$ which captures the relevance of the recommended packages indicating how much the members of $\mathcal{G}$ like the individual items in $\mathcal{P}$.

$$R(\mathcal{G}, \mathcal{P}) = AVG \sum_{u \in \mathcal{G}} \sum_{i \in \mathcal{P}} r(u, i)$$

The second metric is the average item distance $dist(\mathcal{P})$ between items in the package $\mathcal{P}$. This metric indicates how viable the package $\mathcal{P}$ is, i.e., how viable it is for items in that package to be chosen together (items far from each other could be a bad choice in trip planning).

$$dist(\mathcal{P}) = AVG \sum_{i,j \in \mathcal{P}, i \neq j} dist(i,j)$$

6.4 Results

In this section, we discuss the results we obtained from our experiments according to the quality of the recommended packages. We report results by varying the group size, the package size, the number of recommended packages k and finally the type of groups (RAND groups, SIM groups and DIV groups). We note that $R(\mathcal{G}, \mathcal{P})$ and $dist(\mathcal{P})$ are two indicators of package quality, in terms of group preference on items and package viability, respectively. We expect AVG-GREEDY and LM-GREEDY to generate packages with the best $R(\mathcal{G}, \mathcal{P})$, because both baselines are designed to combine items most liked by the groups independently of the relationship among these items, which may lead to non viable packages. A desirable model should generate packages with a similar $R(\mathcal{G}, \mathcal{P})$ to AVG-GREEDY and LM-GREEDY and at the same time find packages with a small $dist(\mathcal{P})$, leading to a high viability of the packages.

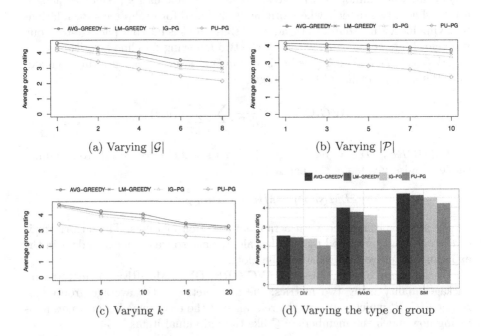

(a) Varying $|\mathcal{G}|$

(b) Varying $|\mathcal{P}|$

(c) Varying k

(d) Varying the type of group

Fig. 2. The average group-item rating

Figure 2 shows the average group-item rating $R(\mathcal{G}, \mathcal{P})$ over the packages recommended by AVG-GREEDY, LM-GREEDY, IG-PG and PU-PG respectively on the Yelp dataset. Since each model recommends a set of top-k packages, we average $R(\mathcal{G}, \mathcal{P})$ over all the recommended packages.

In all settings, AVG-GREEDY and LM-GREEDY perform the best because of their design goal. However, the IG-PG model generates packages that have almost the same group-item rating $R(\mathcal{G}, \mathcal{P})$. On the other hand, PU-PG always performs worst than IG-PG, because of the following drawbacks: (1) It assumes that each user in the group selects the package as a whole regardless of the preferences of other users in the group (2) For the same reason, a user will never select an item that he/she does not like much, i.e., he/she will never compromise for the sake of the group (3) The top-k packages for different users may not overlap to other group members, especially for DIV groups, leaving the packages to be satisfying for single users but not for the whole group.

We examine the effect of different group sizes in Fig. 2(a). We can observe that for all the models, when the group size $|\mathcal{G}|$ rises, $R(\mathcal{G}, \mathcal{P})$ decreases, because it tends to be more difficult to find a consensus over items, between users of large size. Figures 2b and c show that, as expected, when the package $|\mathcal{P}|$ or the number of recommended packages k rises, there is a small decrease on $R(\mathcal{G}, \mathcal{P})$. Another observation (Fig. 2d) is that the group similarity has a direct impact on the relevance of the recommended packages. We notice that for the SIM groups, all models performed slightly the same, as it is easy to find packages where all items satisfy all groups members, even for the PU-PG model. On the other hand, for DIV groups, we notice a decrease on the relevance of the recommended packages which is explained by the fact that, the more heterogeneous are the preferences of the group members, the more it is difficult to find items that will satisfy all the users in the group.

Figure 3 compares the models based on the average distance $dist(\mathcal{P})$ between items in each package. Since each models suggest a set of top-k packages, we also average $dist(\mathcal{P})$ over all recommended packages.

Since AVG-GREEDY and LM-GREEDY ignore relationships between the items, the recommended packages that are suggested may contain items that are far from each other and thus have high $dist(\mathcal{P})$ values. This allows the recommendation of packages that are not viable, which may lead to the users dissatisfaction. On the other hand, IG-PG and PU-PG provide package recommendations with a small $dist(\mathcal{P})$, as their scoring function takes into account the relationship between item during the selection process. We notice that the PU-PG performs worst than IG-PG, because PU-PG fails to find packages with small $dist(\mathcal{P})$ values, which are liked by the group as a whole, but not really liked by some individual group members; hence it performs on $dist(\mathcal{P})$ worse than that of IG-PG. Another interesting observation (Fig. 3d), is that the relative performance among all models is relatively the same regardless of the similarity between groups members (RAND, SIM, DIV).

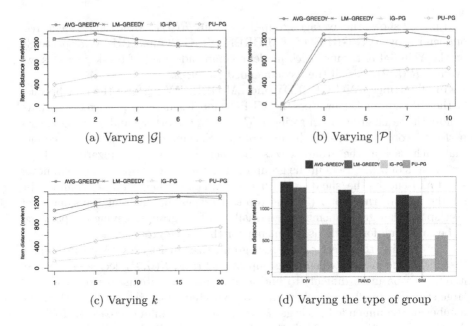

Fig. 3. The average item distance

To sum up, IG-PG performs the best considering $dist(\mathcal{P})$. This model recommends packages that are viable, while being very close to the performance of AVG-GREEDY and LM-GREEDY on the average group-item rating $R(\mathcal{G}, \mathcal{P})$

7   Related Work

7.1   Package Recommendations

One category of previous work deals with recommending a package of items to a single user. In [21], Angel *et al.* are interested in finding the top-k tuples of entities. Examples of entities could be cities, hotels and airlines, while packages are tuples of entities, they query documents using keywords in order to determine entity scores. CourseRank [22] is a project motivated by a course recommendation for helping students planning their academic program at Stanford University. The recommended set of courses must also satisfy constraints (e.g., take two out of a set of five math courses). Similar to our work, each course is associated with a value that is calculated using an underlying recommender engine. Formally, they use the popularity of courses and courses taken by similar students. Given a number of constraints, the system finds a minimal set of courses that satisfy the requirements of the user and has the highest score. The same authors, Parameswaran *et al.*, in [7] extend it with prerequisite constraints, and propose several approximation algorithms that return high-quality course recommendations satisfying all prerequisites. However, these works do not consider constraints for items, while we subject our packages to group-specified constraints, as well as packages viability, which are essential features in the application of trip planning that we consider. Other closely related work is [23] where a framework is proposed to automatically recommend travel itineraries from online user-generated data, like picture uploads using social websites such as Flickr. They formulate the problem of recommending travel itineraries that might be interesting for single users while the travel time is under a given time budget. However, in this work, the value (score) of each item is only determined by the number of times it was mentioned by other users in the social network. In another close work to ours [12], Xie *et al.* explore a method for returning approximate solutions to composite recommendations. The focus of this work is on using a Fagin-style algorithm for variable size packages and proving its optimality. The same authors further develop the idea into a prototype of recommender system for travel planning (CompRec). However, in this work the score of a package is just based on the predicted ratings of a user, ignoring package viability. Furthermore, none of these works accounts for recommending packages to be consumed by a group of users.

7.2   Group Recommendations

Another recent line of work deals with the recommendation of single items to groups of users. The major part of these works investigated the core algorithms used for generating the group recommendations. Different strategies are available and two main approaches have been proposed: Profiles aggregation and recommendations aggregation [4]. The first approach consists of aggregating (combining) the profiles of individual members in order to create a joint *artificial* user profile, which represents all the users in the group. In this way, the group can be treated as a *pseudo* user, and thus standard recommendation techniques can be employed to provide recommendations for the group. The second approach consists in first computing recommendations for each group member separately, and then aggregate them into a single group recommendation list. In this case, first the ranked lists of recommendations for each individual group member are created independently, and then they are aggregated into a final group recommendation list. Inspired by social choice theory [5], numerous aggregation strategies for profiles and recommendations exist. This theory studies how individual preferences could be aggregated to reach a collective consensus. Arrow [5] proved that there is no perfect aggregation strategy, which allows for a variety of different aggregation methods to exist. In [24], Masthoff addressed the problem of how real users aggregate recommendation lists. The goal was primarily to understand how humans aggregate group preferences and hence to select the aggregation method that is best evaluated by the users. This study concluded that Average and Least misery aggregations are those that are human subjects preferred.

MusicFX [25] implements a least misery criterion in order to recommend a list of songs for a group of listeners. *POLYLENS* also aggregates recommendations assuming least misery strategy to extend the well known MovieLens recommendation platform to groups of users. *INTRIGUE* [26] is an interesting hybrid approach that identifies sub-groups among groups (e.g., children, or disabled persons), creates a model for each sub-group, and then merges the sub-group recommendations using an average strategy. Some works (e.g., [27]) use the social relationships between group members to derive group recommendations. Other interesting recent works ([19,20]) considered personal impact weighted models, using the intuition that each user has a different impact during the decision process on the group's selection of items. However, none of these work accounts to recommend packages to groups of users.

8   Conclusion

Motivated by applications of trip planning, we studied in this paper the problem of recommending packages, consisting of sets of items, to groups of users. Each item is associated to one or more categories, which are organized in a tree structure. This allows us to define a semantic similarity measure over items. Each recommended package has to satisfy constraints, that are either *hard*, specified by the group or *soft*, which are desirable but not mandatory constraints. We formalized the problem of top-k package-to-group recommendations and proposed two models (IG-PG and PU-PG), which both incorporate individual ratings given by group members to items, and take into account the aspect of user impact in the selection process. In addition we introduced the notion of package viability. We proposed algorithms that efficiently implement our two models, using a *Produce-and-Choose* paradigm. Our experiments, using a real world dataset, show that the IG-PG model recommends packages of superior quality in terms of package viability, while performing almost the same in terms of group satisfaction, compared to two baseline approaches and PU-PG model.

We plan now to study the fairness problem in package-to-group recommendations. User groups may be heterogeneous with potentially dissimilar tastes. Thus, for a recommended package that is overall good for the group, there might be one or more members that do not like any item in that package, which can lead to a frustration of these users. Therefore, we plan to investigate methods that recommend packages of good quality, while maximizing fairness.

References

1. Resnick, P., Varian, H.R.: Recommender systems. Commun. ACM **40**(3), 56–58 (1997)
2. Lops, P., de Gemmis, M., Semeraro, G.: Content-based recommender systems: state of the art and trends. In: Ricci, F., Rokach, L., Shapira, B., Kantor, P.B. (eds.) Recommender Systems Handbook, pp. 73–105. Springer, Boston (2011). https://doi.org/10.1007/978-0-387-85820-3_3
3. Schafer, J.B., Frankowski, D., Herlocker, J., Sen, S.: Collaborative filtering recommender systems. In: Brusilovsky, P., Kobsa, A., Nejdl, W. (eds.) The Adaptive Web. LNCS, vol. 4321, pp. 291–324. Springer, Heidelberg (2007). https://doi.org/10.1007/978-3-540-72079-9_9

4. Jameson, A., Smyth, B.: Recommendation to groups. In: Brusilovsky, P., Kobsa, A., Nejdl, W. (eds.) The Adaptive Web. LNCS, vol. 4321, pp. 596–627. Springer, Heidelberg (2007). https://doi.org/10.1007/978-3-540-72079-9_20

5. Arrow, K.J.: A difficulty in the concept of social welfare. J. Polit. Econ. **58**(4), 328–346 (1950)

6. Hansen, D.L., Golbeck, J.: Mixing it up: recommending collections of items. In: Proceedings of the SIGCHI Conference on Human Factors in Computing Systems, pp. 1217–1226. ACM (2009)

7. Parameswaran, A., Venetis, P., Garcia-Molina, H.: Recommendation systems with complex constraints: a course recommendation perspective. ACM Trans. Inf. Syst. (TOIS) **29**(4), 20 (2011)

8. Xie, M., Lakshmanan, L.V., Wood, P.T.: CompRec-trip: a composite recommendation system for travel planning. In: 2011 IEEE 27th International Conference on Data Engineering (ICDE), pp. 1352–1355. IEEE (2011)

9. Benouaret, I., Lenne, D.: A package recommendation framework for trip planning activities. In: Proceedings of the 10th ACM Conference on Recommender Systems, pp. 203–206. ACM (2016)

10. Benouaret, I., Lenne, D.: Recommending diverse and personalized travel packages. In: Benslimane, D., Damiani, E., Grosky, W.I., Hameurlain, A., Sheth, A., Wagner, R.R. (eds.) DEXA 2017. LNCS, vol. 10439, pp. 325–339. Springer, Cham (2017). https://doi.org/10.1007/978-3-319-64471-4_26

11. Amer-Yahia, S., Bonchi, F., Castillo, C., Feuerstein, E., Mendez-Diaz, I., Zabala, P.: Composite retrieval of diverse and complementary bundles. IEEE Trans. Knowl. Data Eng. **26**(11), 2662–2675 (2014)

12. Xie, M., Lakshmanan, L.V., Wood, P.T.: Breaking out of the box of recommendations: from items to packages. In: Proceedings of the Fourth ACM Conference on Recommender Systems, pp. 151–158. ACM (2010)

13. Amer-Yahia, S., Roy, S.B., Chawlat, A., Das, G., Yu, C.: Group recommendation: semantics and efficiency. Proc. VLDB Endow. **2**(1), 754–765 (2009)

14. Baltrunas, L., Makcinskas, T., Ricci, F.: Group recommendations with rank aggregation and collaborative filtering. In: Proceedings of the Fourth ACM Conference on Recommender Systems, pp. 119–126. ACM (2010)

15. Stettinger, M.: Choicla: Towards domain-independent decision support for groups of users. In: Proceedings of the 8th ACM Conference on Recommender Systems, pp. 425–428. ACM (2014)

16. Delic, A., et al.: Observing group decision making processes. In: Proceedings of the 10th ACM Conference on Recommender Systems, pp. 147–150. ACM (2016)

17. Sarwar, B., Karypis, G., Konstan, J., Riedl, J.: Item-based collaborative filtering recommendation algorithms. In: Proceedings of the 10th International Conference on World Wide Web, pp. 285–295. ACM (2001)

18. Castillo, L., Armengol, E., Onaindía, E., Sebastiá, L., González-Boticario, J., Rodríguez, A., Fernández, S., Arias, J.D., Borrajo, D.: SAMAP: an user-oriented adaptive system for planning tourist visits. Expert. Syst. Appl. **34**(2), 1318–1332 (2008)

19. Liu, X., Tian, Y., Ye, M., Lee, W.C.: Exploring personal impact for group recommendation. In: Proceedings of the 21st ACM International Conference on Information and Knowledge Management, pp. 674–683. ACM (2012)

20. Yuan, Q., Cong, G., Lin, C.Y.: COM: a generative model for group recommendation. In: Proceedings of the 20th ACM SIGKDD International Conference on Knowledge Discovery and Data Mining, pp. 163–172. ACM (2014)

21. Angel, A., Chaudhuri, S., Das, G., Koudas, N.: Ranking objects based on relationships and fixed associations. In: 12th International Conference on Extending Database Technology, EDBT 2009 (2009)
22. Parameswaran, A.G., Garcia-Molina, H.: Recommendations with prerequisites. In: Proceedings of the Third ACM Conference on Recommender Systems, pp. 353–356. ACM (2009)
23. De Choudhury, M., Feldman, M., Amer-Yahia, S., Golbandi, N., Lempel, R., Yu, C.: Automatic construction of travel itineraries using social breadcrumbs. In: Proceedings of the 21st ACM Conference on Hypertext and Hypermedia, pp. 35–44. ACM (2010)
24. Masthoff, J.: Group recommender systems: combining individual models. In: Ricci, F., Rokach, L., Shapira, B., Kantor, P.B. (eds.) Recommender Systems Handbook, pp. 677–702. Springer, Boston (2011). https://doi.org/10.1007/978-0-387-85820-3_21
25. McCarthy, J.F., Anagnost, T.D.: MUSICFX: an arbiter of group preferences for computer supported collaborative workouts. In: Proceedings of the 1998 ACM Conference on Computer Supported Cooperative Work, pp. 363–372. ACM (1998)
26. Ardissono, L., Goy, A., Petrone, G., Segnan, M., Torasso, P.: INTRIGUE: personalized recommendation of tourist attractions for desktop and hand held devices. Appl. Artif. Intell. **17**(8–9), 687–714 (2003)
27. Li, K., Lu, W., Bhagat, S., Lakshmanan, L.V., Yu, C.: On social event organization. In: Proceedings of the 20th ACM SIGKDD International Conference on Knowledge Discovery and Data Mining, pp. 1206–1215. ACM (2014)

Statistical Relation Cardinality Bounds in Knowledge Bases

Emir Muñoz[1,2(✉)] and Matthias Nickles[2]

[1] Fujitsu Ireland Limited, Dublin, Ireland
emir@emunoz.org
[2] Insight Centre for Data Analytics, National University of Ireland, Galway, Ireland
matthias.nickles@nuigalway.ie

Abstract. There is an increasing number of Semantic Web knowledge bases (KBs) available on the Web, created in academia and industry alike. In this paper, we address the problem of lack of structure in these KBs due to their schema-free nature required for open environments such as the Web. Relation cardinality is an important structural aspect of data that has not received enough attention in the context of KBs. We propose a definition for relation cardinality bounds that can be used to unveil the structure that KBs data naturally exhibit. Information about relation cardinalities such as a person can have two parents and zero or more children, or a book should have one author at least, or a country should have more than two cities can be useful for data users and knowledge engineers when writing queries and reusing or engineering KB systems. Such cardinalities can be declared using OWL and RDF constraint languages as constraints on the usage of properties in the domain of knowledge; however, their declaration is optional and consistency with the instance data is not ensured. We first address the problem of mining relation cardinality bounds by proposing an algorithm that normalises and filters the data to ensure the accuracy and robustness of the mined cardinality bounds. Then we show how these bounds can be used to assess two relevant data quality dimensions: consistency and completeness. Finally, we report that relation cardinality bounds can also be used to expose structural characteristics of a KB by mapping the bounds into a constraint language to declare the actual shape of data.

Keywords: Semantic web · Knowledge bases · RDF
Cardinality bounds · Data shapes

1 Introduction

The generation, use and refinement of Semantic Web knowledge bases (KBs) has been a focus of research attention in both academia and industry alike for many years. Knowledge bases have shown to be of great value in many Artificial Intelligence (AI) application domains, such as natural language processing, robotics, and knowledge representation and reasoning. The Resource

© Springer-Verlag GmbH Germany, part of Springer Nature 2018
A. Hameurlain et al. (Eds.): TLDKS XXXIX, LNCS 11310, pp. 67–97, 2018.
https://doi.org/10.1007/978-3-662-58415-6_3

Description Framework (RDF) is used as a lingua franca for representing knowledge bases in the Semantic Web. RDF is schema-less, which means that it gives freedom to data publishers in describing entities and their relationships as facts. These facts do not need to obey some specified unique global data schema that describes their structure and constraints. In practice, most KBs use multiple—sometimes overlapping—vocabularies [30, 42] (e.g., SKOS, FOAF, DCAT) and avoid to include domain, range or cardinality restrictions because of the contradictions these can generate [11, 32]. To illustrate this, let us consider a KB about countries with two different properties: `gov:hasPopulation` and `dbpedia-owl:populationTotal`.[1] These properties come from different ontologies/vocabularies and represent the same thing, i.e., population of a country; however, their semantics might not be the same or the expected domain/range values could be different. Similarly, two or more labels can be used to refer to the same entity: e.g., http://www.geonames.org/2963597/ireland.html and http://dbpedia.org/resource/Ireland. If one of these ontologies/vocabularies defines a constraint over the resources, this constraint might contradict the definition(s) made somewhere else producing an inconsistency and harming the reasoning capabilities over the data.

The lack of a central schema, in turn, can cause a series of difficulties in the reusability of such data (cf. [5, 16, 18, 21]), where applications might need to rely on the fact that data satisfy a set of constraints. A schema indicates what are valid relations for an entity according to its type, what are the allowed values for the properties, and other constraints that instance data should satisfy. For instance, let us consider a software developer building a user interface (UI) that displays information about countries to end users. For that the developer must query the KB, in other words, she must write queries using the SPARQL query language to fill every property of every country in the UI. However, when building the query to retrieve the population of Ireland she finds that, unexpectedly, there are two mismatching population numbers for that country in the KB. These properties come from different vocabularies and their values, i.e., the populations, do not match. This inconsistency could have been avoided if she knew that some countries have more than one value for the relation population. Such situations—sometimes rare, sometimes very abundant—show a gap between the expected and real structure of KBs. Data users and knowledge engineers would benefit from having an understanding of what information is available to write queries, and to reuse or manage KBs [23, 37].

In data management, *cardinality* is an important property of the structure of data [1, 24]. They constraint the population of relationship types and they help us to understand the meaning of relation and entity types involved. In RDF, a relation cardinality limits the number of values that the relation may have for a given entity. For instance, they can be used to express that a person must have two parents, or a book must have more than one (up to many) authors. This work proposes the automated discovery of relation cardinalities

---

[1] Henceforth, we use prefixes to replace namespaces according to http://prefix.cc/ to shorten the length of URLs.

(i.e., the cardinality of instances of rdf:Property or owl:ObjectProperty or owl:DatatypeProperty) to unveil an important aspect of the structure that KBs naturally exhibit regardless of the presence of any ontology or vocabulary. We show that the aforementioned problem can be partially overcome by mining cardinalities from instance data, and complement previous methods (e.g., [43]) towards the generation of a central schema.

With the motivation of capturing the characteristics that data naturally exhibit, we set the goal of discovering structural patterns using a bottom-up or extensional approach for mining relation cardinality bounds from instance data.

Problem Statement. Recently, RDF constraint languages (e.g., Shape Expressions [31], Shapes Constraint Language (SHACL)[2]) have been defined to satisfy the latent requirements for constraints definition and validation of RDF in real-world use cases. They build upon SPARQL or regular expressions to define so-called *shapes* that perform for RDF the same function as XML Schema, DTD and JSON Schema perform for XML or JSON: delimit the boundaries of instance data. However, their generation is still limited to a manual process.

Relation cardinality in RDF data can be declared using the Web Ontology Language (OWL) or RDF constraint languages (e.g., SHACL, ShEx); however, such declarations are hand-crafted, application-dependent, and not always available, highlighting the need for relation cardinality mining methods. Relation cardinality bounds are hidden data patterns whose discovery unveils an important aspect of the structure of data, and in some cases they might not even be intuitive for data creators [21]. We deal with the problem of identifying relation cardinality bounds such as "a person has two parents" or "a book has minimum one author and maximum eleven". We call this problem the relation cardinality mining problem that we define as follows:

RELATION CARDINALITY MINING PROBLEM
Input: a knowledge base $\mathcal{G}$, and optional context τ.   **Output:** set Σ of relation cardinality bounds that are satisfied by $\mathcal{G}$.

It is important to notice that unlike traditional databases, RDF and OWL assume the open-world semantics (OWA), and absence of the unique name assumption (nUNA). This makes the problem of extracting relation cardinality more complex than a naïve application of SPARQL queries using the COUNT operator. Take as example the constraint "a person must have two parents": if the data contain an entity of type Person with only one parent, *this does not cause a logical inconsistency, it just means it is incomplete, and in RDF/OWL incomplete is different from inconsistent.* To deal with these specifics, we propose a method which tackles two important challenges: (1) *KB equality normalisation*, meaning that we must deal with owl:sameAs (or equivalent) axioms representing equality between entities to discover accurate cardinalities, and (2) *outliers filtering*, where we should account for the probability of noise in the data to discover robust cardinalities. By doing so, the result of this work can provide users

---

[2] https://www.w3.org/TR/shacl/ (accessed on February 13, 2017).

with 'shapes' of data that serve them to analyse completeness and consistency, and thus, contribute towards higher levels of quality in KBs [12,26,37].

Contributions. This paper is an extension of [22]: we added a much more thorough description of our method to mine relation cardinality bounds from KBs. Moreover, we show the direct relationship between cardinality bounds and the Shape Expressions (ShEx) language for RDF validation, and describe how the former can be mapped to the latter to build schemas for KBs. We also perform a new experiment focused on the inference of schemas using two new datasets compared to [22].

The present article brings the following technical contributions:

(a) We present an algorithm to extract accurate and robust cardinality bounds for relations in a KB. Our approach deals with equality axioms for entities and identifies outliers in the distribution of cardinalities.
(b) We propose to use the relation cardinality bounds of a KB to assess its completeness and consistency. The discovered relation cardinality bounds are a good indicator of which relations (and entities) are incomplete or inconsistent compared with the median value in the bounds.
(c) We present a mapping of relation cardinality bounds to ShEx data shapes. Our relation cardinality bounds are an indicator of the structural characteristics that data naturally exhibit, thus they can be used to generate simple but insightful data shapes for a KB.

Organisation. The rest of this article is organised as follows. We start reviewing the related work about cardinality, consistency, completeness and schema discovery in KBs in Sect. 2. In Sect. 3, we introduce preliminaries about RDF and Shape Expressions language for validation or RDF data. Then, we provide definitions, semantics and examples for understanding relation cardinality bounds in KBs in Sect. 4. In Sect. 5, we propose an algorithm for mining relation cardinality bounds in an accurate and robust manner, and describe two different ways to implement the normalisation step. The serialisation of cardinality bounds as normative shapes for KBs using ShEx is detailed in Sect. 6. Finally, we evaluate the usage of the mined relation cardinality bounds over different datasets in Sect. 7, and conclude in Sect. 8 with the limitations and future directions.

2　Related Work

We discuss the related work along three main topics.

Cardinality Constraints/Bounds. Cardinality constraints in RDF have been defined for data validation in languages such as OWL [20], Shape Expressions (ShEx) [31], OSLC Resource Shapes [35], and Dublin Core Description Set Profiles (DSP)[3]. In UML and ER diagrams, cardinality bounds are called multiplicities. Yet, UML and ER are not meant to model graph data such as the

---

[3] http://dublincore.org/documents/dc-dsp/.

one in KBs, let alone to describe their structure in a normative way. OSLC integrity constraints include cardinality of relations which are more similar to UML cardinality for associations (i.e., exactly-one, one-or-many, zero-or-many, and zero-or-one). However, the expressivity of OSLC is limited compared to the definitions proposed in OWL[4], DSP, ShEx, SHACL, and Stardog ICV[5]. All these last languages define flexible boundaries for cardinality constraints: a lower bound in $\mathbb{N}$, and an upper bound in $\mathbb{N} \cup \{\infty\}$. SPIN[6] Modelling Vocabulary is yet another language based on SPARQL to specify rules and logical constraints, including cardinality. In this work, we use ShEx language to express the cardinalities in a human-friendly syntax. Due to the bottom-up approach taken here we do not refer to our cardinalities as constraints but as *bounds* [22] that do not constrain the data but indicate soft properties boundaries. They can—and should—be considered as constraints (i.e. hard bounds) only after a user assessment and application over a dataset. Despite this, our work builds upon existing approaches for cardinality constraints in RDF and other data models such as XML [8] and Entity-Relationship [17,40].

Consistency and Completeness in RDF Graphs. Consistency is a relevant dimension of data quality, and many researchers have investigated the checking and handling of inconsistencies in RDF. However, to the best of our knowledge, this is the first work[7] focused on the extraction and analysis of cardinalities to detect inconsistencies in RDF through the application of outlier detection techniques. The concept of outliers or anomaly detection is defined as "finding patterns in data that do not conform to the expected normal behaviour" (a survey related to the subject is proposed in [6]). Under the assumption that KBs are likely to be noisy and incomplete [28], there exist several approaches that aim to enhance or refine their quality and completeness (see Paulheim [26] for a recent survey). Among the most relevant to our work are: [44] which applies unsupervised numerical outlier detection methods to DBpedia for detecting wrong values that are used as literal objects of a (`owl:DatatypeProperty`) property; and [9] that builds upon [44] to identify sub-populations of instances where the outlier detection works more accurately by using external datasets accessible from the `owl:sameAs` links. Our work fundamentally differs from these works in that we focus on cardinality or multiplicity of properties while they focus their attention in finding wrong or missing property values.

RDF Schema Discovery. Our work falls into the broader area of schema discovery. Völker and Niepert in [43] introduce a statistical approach where association rule mining is used to generate OWL ontologies from RDF data, but consider cardinality restrictions (mentioning upper bounds) only as future work. Similarly, in [14] the authors extract type definitions described by profiles, i.e.,

---

[4] OWL allows the expression of cardinalities through the `minCardinality`, `maxCardinality`, and `cardinality` restrictions.

[5] http://docs.stardog.com/icv/icv-specification.html.

[6] http://spinrdf.org/.

[7] This work extends our previous work in [22].

property vector where each property is associated to a probability. A further analysis of semantic and hierarchical links between types is carried to extract a global schema. A hierarchical organisation of entity types (classes) is extracted in [7] using clustering analysis at instance-level grouping together entities with similar sets of properties. Their hierarchical organisation does not take into account any notion of cardinality. They serialise the inferred schema—modelled using UML—back to RDF triples using a vocabulary defined by themselves, whereas in our work we decide to serialise the inferred schema to ShEx, which is more widely used to date and supported by several validation tools. Considering that most KBs are generated from semi- or un-structured data, [41] analysed the case of DBpedia enrichment with axioms identified during the extraction process from Wikipedia. Such axioms are identified with methods from Inductive Logic Programming (ILP), like in [43]. Despite their bottom-up or extensional approach, similar to ours, such works aim to build or enrich ontologies with missing relations, not considering any notion of cardinality nor the use of these to analyse data quality (e.g., completeness, consistency). A related approach to detect cardinality in KBs is presented by Rivero et al. in [32] and uses straight-forward SPARQL 1.1 queries to discover ontological models. Such models include types and properties, sub-types, domain, range, and minimum cardinalities of these properties. However, the approach presented in [32] is not able to deal with the semantics of data: both the existence of equality axioms and outliers or errors in the data are ignored. For these reasons, our work is orthogonal and complementary to all the aforementioned works.

3   Preliminaries

This section provides some introductory notions regarding RDF and its semantics, and the role of Shape Expressions in the RDF data validation.

RDF Model. We define a Semantic Web knowledge base following [3]. Let $\mathcal{R}$ be the set of *entities*, $\mathcal{B}$ the set of *blank nodes*, $\mathcal{P}$ the set of *predicates*, and $\mathcal{L}$ the set of *literals*. A finite *knowledge base* $\mathcal{G}$ is a set of *triples*, say $\mathcal{G} = \{(s, p, o)\}_{i=1}^{m}$, where each $(s, p, o) \in (\mathcal{R} \cup \mathcal{B}) \times \mathcal{P} \times (\mathcal{R} \cup \mathcal{B} \cup \mathcal{L})$. In each (s, p, o), s denotes the *subject*, p denotes the *relation* predicate, and o denotes the *object*. The sets $\mathcal{R}$, $\mathcal{B}$, $\mathcal{P}$ and $\mathcal{L}$ are infinite and pair-wise disjoint. We denote by $\mathscr{G}$ the infinite set of all possible knowledge bases.

Additionally, we define the functions $\mathbf{pred}(\mathcal{G}, \tau) = \{p \mid \exists s, o \ (s, p, o) \in \mathcal{G} \wedge (s, \mathtt{rdf{:}type}, \tau) \in \mathcal{G}\}$ that returns a set of predicates appearing with instances of entity type τ; and $\mathbf{triples}(\mathcal{G}, \Sigma, p) = \{(s, p, o) \mid \exists o \ (s, p, o) \in \mathcal{G} \wedge s \in \Sigma\}$ that returns the triples in $\mathcal{G}$ with subject s in an input set Σ and relation p. Likewise, we define $\mathbf{sameAsPairs}(\mathcal{G}) = \{(s, o) \mid \exists \ s, o \ (s, \mathtt{owl{:}sameAs}, o) \in \mathcal{G}\}$ as the function that returns all pairs of entities defined as equal but with different naming. Note that the predicate $\mathtt{owl{:}sameAs}$ is used in most cases to represent equality, but other predicates (e.g. $\mathtt{same\_as}$) could also be used for the same purpose.

UNA 2.0. The *unique name assumption* (UNA) is a simplifying assumption made in some ontology languages and description logics. It entails that two different names always refer to different entities in the world [34]. On the one hand, OWL default semantics does not adopt the UNA, thus two different constants can refer to the same individual. This is a desirable behaviour in an environment such as the Web, where reaching an agreement for labelling entities is unfeasible [13]. On the other hand, validation checking approaches in RDF usually adopt a *closed-world assumption* (CWA) with UNA, i.e., inferring a statement to be false on the basis of failure to prove it, and if two entities are named differently they are assumed to be different entities. To deal with this, SHACL defines the so-called *UNA 2.0* which is a simple workaround where all entities are treated as different, unless explicitly stated otherwise by `owl:sameAs` (or equivalent) relation. From a practical point of view, it is a desirable feature for mining algorithms to consider the semantics of knowledge bases avoiding misinterpretations of the data. Figure 1 (left) shows an example where the adoption of normal UNA will lead to a count of five different entities (i.e., `ex:A`, `ex:B`, `ex:C`, `ex:D`, `ex:E`), and `ex:A` would have cardinality one for the property `ex:p1`. While adopting UNA 2.0 (Fig. 1, right) the counts change: now we have four different entities (i.e., `ex:A`, `ex:B`, `ex:C`, `ex:E`), and the cardinality of `ex:p1` in `ex:A` is 2. Here, we call *rewriting* the process of applying UNA 2.0 to a previously unnormalised KB.

Since the cardinality of a property is severely affected when sameAs-axioms are not considered, hereafter, we adopt UNA 2.0 and satisfy this requirement, allowing us to correctly interpret a KB semantics.

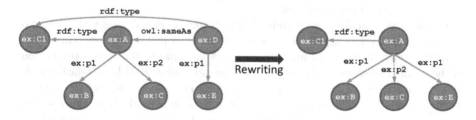

Fig. 1. Example of UNA 2.0 as defined in SHACL.

Shape Expressions. Shape Expressions (ShEx)[8] is intended to be an RDF constraint language, which allows to describe the structure or "shape" of RDF graphs (i.e., knowledge bases) [31]. A *ShEx schema* is a collection of expressions that describe the constraints that a given RDF graph should meet to be considered valid. It contains the allowed values for the predicates of an entity, directionality of the relations, and cardinality constraints for the number of allowed occurrences of a relation. As an example, consider the ShEx schema in Fig. 2 that constrains instances of a `ex:Customer` entity type as follows: a single

---

[8] Shape Expressions (ShEx) Primer: http://shex.io/shex-primer/.

first name of type literal; one to three last names of type literal; an age between 18 and 99 years; zero or more ('$*$' in regular expressions) e-mail addresses for contact; and the customer must be referenced in at least one ('$?$' in regular expressions) payment report (defined by another ShEx shape indicated by the '$@$' symbol). Notice that a ShEx schema can be serialised in three different ways, namely, **ShExC** for Shape Expressions Compact Syntax, **ShExJ** for JSON-LD (JavasSript) syntax, and **ShExR** for RDF Turtle syntax. In this work, we use ShExC because it is the most human-friendly syntax among the options. For more details, we refer readers to the ShEx Primer.

```
1  ex:CustomerShape {
2    foaf:firstName xsd:string {1} ;
3    foaf:lastName xsd:string {1, 3} ;
4    foaf:age xsd:integer MinInclusive 18 MaxInclusive 99 ;
5    foaf:mbox IRI * ;
6    ^ex:payment @ex:PaymentReportShape ?
7  }
```

Fig. 2. Example ShEx schema for a `ex:Customer` entity type.

4   Relation Cardinality Bounds

In this section, we introduce a definition of relation cardinality bounds for KBs that generalises the semantics of the definitions discussed in Sect. 3. Some of these definitions were already introduced in our previous work [22] and repeated here for the sake of completeness.

Cardinality (also known as multiplicity) covers a critical aspect of relational data modelling, referring to the number of times an instance of one entity can be related with instances of another entity. A *cardinality bound* is a restriction on the number of elements in the relation. In particular, KBs have relationships between entities (of a given type) connected by relation types, and we want to specify bounds for such relationships. For example, we would like to express that a drug has only one molecular formula, but can be associated to a finite set (of known or unknown size, the latter denoted as ∞) of adverse drug reactions.

Definition 1. *The predicate count of p with respect to s, denoted* $\mathbf{count}(p, s)$, *is defined as the number of triples in $\mathcal{G}$ with s as subject and p as relation:*

$$\mathbf{count}(p, s) = |\{(s, p, o) \mid \exists o, (s, p, o) \in \mathcal{G}\}| .$$

Considering that the **count** function counts the number of *different* objects appearing with a given subject–relation pair, it is easy to see how the output of the function is directly affected by the rewriting of the KB $\mathcal{G}$.

To illustrate the definition above, let us consider a KB about books and say that the book *Foundations of Databases* has three authors and one publisher. Thus, the **count**(*author, Foundations of Databases*) = 3 and **count**(*publisher, Foundations of Databases*) = 1.

Definition 2. *A relation cardinality bound in Semantic Web knowledge bases restricts the number of relation values related with an entity in a given context. Such context could be a particular entity type or the whole KB. Formally, a cardinality bound φ is an expression of the form $card(P, \tau) = (min, max)$, where $P \subseteq \mathcal{P}$, τ is a context entity type, and where $min \in \mathbb{N}$ and $max \in \mathbb{N} \cup \{\infty\}$ with $min \leq max$. Here $|P|$ denotes the number of properties in φ, min is called the* lower bound, *and max the* upper bound *of φ. If τ is defined ($\tau \neq \varepsilon$), we say that φ is* qualified; *otherwise we say that φ is* unqualified.

The semantics of this definition of relation cardinality bounds limits the maximum and minimum counts that a given set of relations can have in a given context—as in SHACL, DSP, ICV and OWL. The lower bound of a cardinality may take on values in $\mathbb{N}$, whilst upper bounds can be ∞ to represent that there is an unknown upper limit. In fact, each RDF constraint language has different default values for the minimum and maximum cardinalities. For instance, ShEx assigns a default cardinality of one to a predicate appearing in a shape without any explicit cardinality. On the other hand, SHACL assumes a lower cardinality of zero and upper of ∞. A constraint with such default values (i.e., zero and ∞) will always be satisfied by the data, thus it may be omitted from some data shapes leaving a gap in the explicitness of the shape and the structure of data. We deal with that gap in Sect. 5, where we introduce an algorithm to mine relation cardinality bounds from KBs.

An unqualified bound is independent of a type/context, i.e., it holds for a set of relations independent of its context, whereas a qualified bound holds only for a set of relations in combination with subject entities of a same given type. Herein, we focus on qualified constraints given their interestingness and relevance for structural analyses of KBs.

Definition 3. *Consider a KB $\mathcal{G}$. We say that φ is a cardinality bound in $\mathcal{G}$ or $\mathcal{G}$ satisfies φ for a set of relations $P_\varphi \subseteq \mathcal{P}$, a lower bound min_φ, and upper bound max_φ defined in φ, denoted by $\mathcal{G} \models \varphi$, iff*

$$\forall s \in (\mathcal{R} \cup \mathcal{B}), \ p \in P_\varphi \ (min_\varphi \leq \textbf{count}(p, s) \leq max_\varphi).$$

If φ is qualified to τ then to satisfy φ, $\mathcal{G}$ also needs to satisfy the condition that $\forall s \in (\mathcal{R} \cup \mathcal{B}) \ (s, \text{rdf} : \text{type}, \tau) \in \mathcal{G}$.

To illustrate the definitions above, let us consider again the KB about books. We can encode the constraint "a book can have between one and ten authors" using $card(\{author\}, Book) = (1, 10)$. However, if a new book is added to $\mathcal{G}$ with eleven (or zero) authors, then this book violates this constraint, denoted by $\mathcal{G} \not\models \varphi$.

Although the mining approach that we will present in Sect. 5 is able to compute an upper bound cardinality, such limit is uncertain when considering RDF's open world assumption (OWA). For instance, even when the data show that an entity of type Person has at most two children, this might be wrong when considering other unseen same type instances. More certain cardinality bounds can be mined from reliable or complete KBs, which are rarely present on the Web and usually existent within specific domains. Therefore, we refer to relation cardinality bounds as *patterns* when they are automatically extracted from KBs, and as *constraints* when normatively assessed by a user and applied to restrict a KB.

```
1  SELECT $this
2  WHERE {
3          $this $PROPERTY ?value .
4        }
5  } GROUP BY $this
6  HAVING (COUNT(?value) <= $minCount)
```

Fig. 3. SPARQL 1.1 definition of a minimum cardinality constraint.

In practice, cardinality bounds can be used to validate KBs using SPARQL 1.1 queries. For instance, Fig. 3 shows the SPARQL query with aggregation proposed to validate a lower bound min_φ (`$minCount`). The query represents restrictions on the number of values, `?value` nodes, that the `$this` node may have for the given property. A validation result must be produced if the number of value nodes is more than `$minCount`, thus indicating that the data do not conform to the shape. Likewise, to validate an upper bound (max_φ) restriction for a property, we change the `HAVING` condition to '>=', and return a validation result if the number of values is less than `$maxCount`. Note that SHACL, ShEx, and other constraint languages only allow the definition of one condition at a time per property. Therefore, to validate our cardinalities with multiple properties, one must apply an SPARQL 1.1 query like the one in Fig. 3 independently for each entity and property pair with a single bound (upper or lower). In Sect. 5 we will show how a single, but much more complex, SPARQL 1.1 query can be used to extract both minimum and maximum bounds at once.

Example 1. *The following expressions define cardinality bounds for different entity types in different domains. Abusing of notation, we may write $card(p, \tau) = (min, max)$ when the constraint applies to a single relation.*

1. $card(\{\texttt{mondial:name}, \texttt{mondial:elevation}\}, \texttt{mondial:Volcano}) = (1, 1)$,
2. $card(\texttt{mondial:hasCity}, \texttt{mondial:Country}) = (1, \infty)$,
3. $card(\texttt{dcterms:contributor}, \texttt{bibo:Book}) = (0, \infty)$,
4. $card(\texttt{dcterms:language}, \texttt{bibo:Book}) = (1, 2)$.
5. $card(\{\texttt{lmdb:editor}, \texttt{lmdb:filmid}\}, \texttt{lmdb:Film}) = (1, 9)$.

As suggested in the previous examples, when the upper bound is unclear we can use ∞ in the cardinality bound to express that uncertainty.

5   Mining Cardinality Bounds

In this section, we introduce an algorithm for mining relation cardinality patterns from KBs. We also present two different implementations: one based on SPARQL

1.1 that uses a graph database approach to normalise and extract cardinalities; and another based on Apache Spark[9] that applies a MapReduce or divide-and-conquer strategy to divide the data and run the normalisation step in parallel.

5.1   Algorithm

We present Algorithm 1 as an efficient and domain-agnostic solution to mine accurate and robust cardinality patterns from any KB. This algorithm is designed to mine qualified cardinalities, i.e., a context type is given; however, it can be easily adapted to mine unqualified cardinalities. From a data quality perspective, it is desirable that the mined relation cardinality bounds (cf. Sect. 4) are accurate and robust. Algorithm 1 outputs a set of relation cardinality patterns, which are called *accurate* because it considers the semantics of equality axioms (expressed by owl : sameAs and equivalent predicates), and *robust* because we perform an outliers detection and filtering over noisy cardinality counts.

Our mining algorithm (cf. Algorithm 1) has three main steps:

(1) **KB normalisation:** represented by the function **normalise** : $\mathcal{G} \mapsto \mathcal{G}$, receives an unnormalised KB (with possibly multiple equal entities) and applies an on-the-fly (in memory) rewriting process to consider the semantics of the relation owl : sameAs (or other equivalent relation). We address the normalisation by querying all equal entities and building graphs connecting these nodes, which builds so-called cliques where all equal nodes are connected to each other. The cliques are used to normalise the whole KB with replacements. This step could be considered optional in cases where users want information about unnormalised bounds—at the cost of accuracy.

(2) **Cardinalities extraction:** performed by the function **cardPatterns** : $\mathcal{G} \mapsto \mathcal{E} \times \mathbb{N}$, it is called to retrieve (entity = cardinality) pairs from the passed set of triples in the context of a given relation. The cardinalities for all relations are stored in a map, which either filtered from noisy values or returned directly to extract the constraints.

(3) **Outliers filtering:** represented by the **filterOutliers** : $\mathcal{E} \times \mathbb{N} \mapsto \mathbb{N} \cup \emptyset$ function, receives a map of (entity = cardinality) pairs and applies grouping and unsupervised univariate statistical methods to identify and remove noisy or outside of a range values to ensure robustness. It returns an empty set when the (entity = cardinality) pair is filtered out and the cardinality otherwise. Similarly to normalisation, this step could be considered optional at the cost of robustness though.

Next, we present an example for the application of Algorithm 1, and describe each of its part in more details in Sects. 5.2 to 5.4.

Example 2. *Let us consider a KB with entities* ex : s1, ex : s2, *and* ex : s3, *and* properties ex : p1 and ex : p2. *First, we apply the* **normalise** *function that replaces entities by one representative element equivalence type induced according to the*

---

[9] http://spark.apache.org/ (version 2.1.0).

Algorithm 1. Extraction of cardinality bounds

Input: a knowledge base $\mathcal{G}$; and a context τ
Output: a set Σ of cardinality bounds
1: $\mathcal{G}' \leftarrow normalise(\mathcal{G})$
2: $E \leftarrow \{s \mid (s, \, \texttt{rdf:type}, \tau) \in \mathcal{G}'\}$
3: $P \leftarrow pred(\mathcal{G}', \tau)$ ◁ *predicates for entities of type* τ
4: **for all** $p \in P$ **do**
5:     $\mathcal{G}'' \leftarrow triples(\mathcal{G}', E, p)$ ◁ *triples with entity type* τ *and predicate p*
6:     $M \langle u, v \rangle \leftarrow cardPatterns(\mathcal{G}'')$ ◁ *map: u is an entity, and v a cardinality*
7:     $\Theta \leftarrow filterOutliers(M)$ ◁ *set of inliers*
8:     $\Sigma.add(card(\{p\}, \tau) = (min(\Theta), \, max(\Theta)))$
9: **end for**

sameAs-cliques. Then, for each property we extract the (entity, cardinality) pairs using the function **cardPatterns**. *For instance, we obtain* $\{$ex:s1 = 1, ex:s2 = $1, $ex:s3 = 2$\}$ *for property* ex:p1, *and* $\{$ex:s1 = 3, ex:s2 = 25, ex:s3 = 3$\}$ *for property* ex:p2. *Next, the function* **filterOutliers** *tries to identify outliers and determines that there are no outliers for property* ex:p1, *but that a cardinality of 25 is an outlier for* ex:p2. *Thus, 25 is removed from the patterns leaving* $\{$ex:s1 = 3, ex:s3 = 3$\}$ *as robust cardinalities for* ex:p2. *Finally, the cardinality bounds (min, max) are extracted from the remaining inlier cardinalities by using simple* **min** *and* **max** *functions.*

5.2   Knowledge Bases Equality Normalisation

Knowledge bases contain different types of axioms, being owl:sameAs and equivalent-semantic relations the most important when computing cardinalities. Regardless of the approach, by not considering these axioms a method loses its accuracy and cannot ensure that the relation cardinality bounds are consistent with the data and domain of knowledge. Unlike [32], we perform an on-the-fly normalisation of the graph to capture the semantics of sameAs-axioms without having to modify the underlying data. A naïve approach using a SPARQL query with COUNT operator will wrongly return two instances of ex:C1 instead of the expected count of 1 for the example in Fig. 1 (left).

Table 1. An axiomatisation for reduction on equality

$(s, p, o) \land (s', p', o') \land$ $(s', \texttt{owl:sameAs}, s)$	$(s, p, o) \land (s', p', o') \land$ $(o', \texttt{owl:sameAs}, o)$
$(s, p, o), (s, p', o')$ (subject-equality)	$(s, p, o), (s', p', o)$ (object-equality)

To overcome this issue we propose an axiomatisation with two rules (cf. Table 1), namely, *subject-equality* and *object-equality*. The axiomatisation imposes that duplicated elements are replaced by a representative element of equivalent type induced by `owl:sameAs` or similar properties. This normalisation can be done replacing the underlying data [19] or on-the-fly (without modification) when needed. However, if the underlying data is modified, the links to other KBs stated by the sameAs-axioms are overwritten and lost.

Instead, here we follow an on-the-fly rewrite (Line 1 of Algorithm 1) which performs the modifications in memory. A similar approach was previously used by Schenner et al. in [36]. In this rewriting approach, all the so-called *sameAs-cliques* are replaced by a selected representative entity [19]. sameAs-cliques are built by connecting the triples returned by the ***sameAsPairs*** function, generating complete graphs[10] with entities as nodes all of which are equal to each other. In each clique, all nodes (entities) are connected to each other, and a representative node can be chosen randomly. The representative is then used to rewrite the KB: replacing all the appearances of the equal entities in the clique by the representative everywhere in the KB. Note that for doing an on-the-fly rewriting, it is assumed that all cliques either fit in memory or are stored in an external fast-access index. This could be considered a limitation of our approach; however, in our experiments we did not find any case where these cliques did not fit in main memory.

In practice, the axiomatisation of Table 1 can be implemented on-the-fly either by using SPARQL 1.1 or programmatically. Next, we briefly introduce these two options:

SPARQL Rewriting. The SPARQL query language has the limitation that it is unaware of the special semantics of `owl:sameAs` when evaluating triple patterns. This is a serious problem, especially when using the language to count the cardinality of properties. However, one can make use of several constructs in the language to overcome such limitation. We make use of a nested SPARQL 1.1 query with sub-selects [29] as shown in Fig. 4, which contains three sub-queries (`SQ-1`, `SQ-2` and `SQ-3`) with a wrapping query that aims to obtain the (entity = cardinality) pairs for a given property and entity type (Line 3). Under `SQ-1` are `SQ-2` and `SQ-3`, which perform the sameAs-clique generations for subjects and objects, respectively. Sub-query `SQ-2` implements the subject-equality rule, whereas sub-query `SQ-3` implements the object-equality rule. During the clique generation, a graph search to find equal entities is performed in all directions from a starting node using the property path `(owl:sameAs|^owl:sameAs)*`, which is a complex and resource-demanding query (check [2,15] for a more detailed study on property paths in RDF). For each clique found, a representative is selected, and all "clone" entities are rewritten with the representative.

Clearly, the query used here is complex and resource demanding when executed with medium and large KBs. In terms of complexity, the evaluation of SPARQL queries can be solved in $O(|P| \cdot |\mathcal{G}|)$, where $|P|$ is the number of graph patterns in the query and $|\mathcal{G}|$ the number of triples in the knowledge base [3]. SPARQL is known to be in PSPACE-complete in general [3]. Hence, we also

---

[10] Any complete graph is its own maximal clique.

propose a more efficient and faster solution that works outside of a SPARQL endpoint.

```
 1 PREFIX rdf: <http://www.w3.org/1999/02/22-rdf-syntax-ns#>
 2 PREFIX owl: <http://www.w3.org/2002/07/owl#>
 3 SELECT ?first_subj (COUNT(DISTINCT ?first_obj) AS ?nbValues) WHERE {
 4     { SELECT DISTINCT ?first_subj ?first_obj WHERE {              % (SQ-1)
 5         ?subj $property ?obj .
 6         { SELECT ?subj ?first_subj WHERE {                       % (SQ-2)
 7             ?subj a $type .
 8             ?subj ((owl:sameAs|^owl:sameAs)*) ?first_subj .
 9             ?notfirst ((owl:sameAs|^owl:sameAs)*) ?first_subj .
10             FILTER (STR(?notfirst) < STR(?first_subj))
11         } FILTER(!BOUND(?notfirst))
12     }}
13     { SELECT ?obj ?first_obj WHERE {                             % (SQ-3)
14         ?obj ((owl:sameAs|^owl:sameAs)*) ?first_obj .
15         ?notfirst ((owl:sameAs|^owl:sameAs)*) ?first_obj .
16         FILTER (STR(?notfirst) < STR(?first_obj))
17     } FILTER(!BOUND(?notfirst))
18     }}
19   }}
20 } GROUP BY ?first_subj
```

Fig. 4. Query the cardinality of a property for every entity of a given type.

Programmatic Rewriting. Because of the complexity of the SPARQL solution, we propose a second rewriting approach that promises to be more time- and space-efficient. We thus frame the extraction of relation cardinality patterns as the well-known words count problem from linguistics. This problem has been one of the first to be addressed by modern parallelisation paradigms and frameworks. Thus, we can easily parallelise the algorithm using frameworks such as Apache Spark. By using Spark and the `filter` and `map` operations, we implemented a parallel and efficient rewrite, where the sameAs-cliques are generated and used to normalise the KB triple by triple (see Fig. 5, left). We generate the sameAs-cliques as follows: for each $(s, owl{:}sameAs, o)$ triple we lexically compare s and o and select the minimum (say s), which becomes the *representative*; add a mapping from the representative entity to the other (say from s to o); if the non-representative (say o) in this axiom was the representative of other entities, then we update their mappings in cascade with the newly found representative (say s). We then apply a `map` operation over each initial triple in $\mathcal{G}$ and rewrite it according to the sameAs-cliques to obtain $\mathcal{G}'$ in $O(1)$.

5.3   Detection of Cardinality Patterns

After the normalisation step is finished, cardinalities can be collected for each relation (Algorithm 1, Line 6). This ensures their accuracy, which is a major difference w.r.t. previous approaches such as Rivero et al. [32]. In the SPARQL-based approach, Fig. 4 shows a query which performs both the normalisation of $\mathcal{G}$ and the detection of cardinality patterns in one place. However, complex

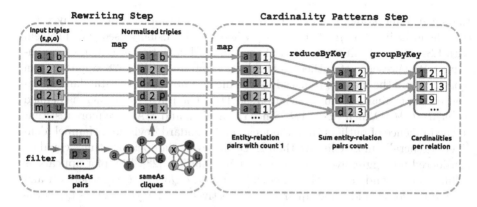

Fig. 5. Cardinality patterns extraction using Apache Spark.

SPARQL queries are hard to evaluate and optimise, making this approach very inefficient and poorly scalable [38]. On the other hand, the Spark-based approach can make use of multiple machines to scale and process large KBs in splits. We show a comparison between the two approaches in Sect. 7. Regardless of the rewriting approach, the output of the cardinality extraction is a map of (entity = cardinality) pairs for a given relation and entity type.

Users could take these cardinalities as patterns at this stage; however, several works have shown that KBs frequently contain noise and outliers (e.g., [12, 26, 27]). To tackle this, we carry out a filtering of outliers from the cardinalities, which is described in the next section.

5.4 Outlier Detection and Filtering

Considering the adverse effects that outliers could cause in the method described so far, we now present techniques that can be used to detect and remove cardinality outliers from KBs (cf. Algorithm 1, Line 7). Several supervised and unsupervised approaches can be used for the detection of outliers in numerical data (see [28] for details); however, we did not find any labelled dataset for valid cardinality values. Therefore, we only consider unsupervised approaches for univariate data. We address the detection of outliers in a sequence of numbers as a statistical problem. Statistical outlier detection methods define rules for identifying points that do not respect the nominal behaviour of the data (i.e., appear to be anomalous) but they cannot explain the reason(s). Usually, the interpretation of outliers depends on the domain of knowledge and the nature of the data, thus, there are no universal rules for that. Interestingly, outlier detection approaches determine a lower and upper bound on the range of data, similarly to the semantics of a cardinality bound.

We studied three of the most commonly used methods for identifying outliers in univariate data: ESD, Hampel and Bloxplot. The *extreme studentized deviation* (ESD) identifier [33] is one of the most popular approaches. It computes the mean μ and standard deviation σ values and considers as outlier any value outside of the interval $[\mu - t \cdot \sigma, \mu + t \cdot \sigma]$, where $t = 3$ is usually used. The problem with ESD is that both the mean and the standard deviation are themselves sensitive to the presence of outliers in the data. *Hampel* identifier [28] appears as an option, where the mean is replaced by the median *med*, and the standard deviation by the median absolute deviation (MAD). The range for outliers is defined as $[med - t \cdot MAD, med + t \cdot MAD]$. Since the median and MAD are more resistant to the influence of outliers than the mean and standard deviation, Hampel identifier is generally more effective than ESD. However, Hampel sometimes could be considered too aggressive, declaring too many outliers [28]. Box plot appears as a third option, and defines the range: $[Q1 - c \cdot IQD, Q3 + c \cdot IQD]$, where $Q1$ and $Q3$ are the lower and upper quartiles, respectively, and $IQD = Q3 - Q1$ is the interquartile distance—a measure similar to the standard deviation. The parameter c is similar to t in Hampel and ESD, and is commonly set to $c = 1.5$. Box plot is better suited for distributions that are moderately asymmetric, because it does not depend on an estimated "centre" of the data. Thus, in our evaluation we use the box plot rule to determine cardinality outliers.

6   Mapping Cardinality Bounds to ShEx

We already mentioned that the discovery of cardinality bounds captures and unveils the structure that data naturally exhibit. Following that intuition, it is straightforward to use the mined cardinality bounds to build ShEx schemas (as the one shown in Fig. 2) to represent the structure of a knowledge base.

When compared to the example in Fig. 2, the shapes we aim to build do not consider datatypes or domain and range values, thus, we will not specify whether a property range is `xsd:string`, `xsd:integer` or `IRI`, for example. Here, we focus on the capability of ShEx schemas to define the graph topology of knowledge bases. Also, we will not consider recursive or external references to other shapes, meaning that we will not have ShEx expressions such as `'^ex:payment @ex:PaymentReportShape ?'` (see Line 6 in Fig. 2), where a second ShEx shape, `@ex:PaymentReportShape`, is referenced. This is out of the scope of this article, and could be considered as a future research direction. Next, we focus mainly on generating ShEx schemas that state the cardinality bounds of entity types.

We follow Boneva et al. [4] for describing a subset of ShEx shape schemas. A shapes schema $\mathcal{S}$ defines a set of named shapes, where a shape is a description of the graph structure that can be visited starting from a particular node. Shapes are expressed by shape expressions, can use (Boolean combinations of) other shapes and recursion is allowed. Formally, a *shapes schema* $\mathcal{S}$ is a pair $(\mathcal{L}, \boldsymbol{def})$, where $\mathcal{L}$ denotes a set of *shape labels* used as names of shapes and $\boldsymbol{def}$ is a function that maps a shape label with a shape expression. $L \rightarrow S$ is used as a short for $\boldsymbol{def}(L) = S$, where a shape label L is associated with a shape expression S. A *shape expression* is a Boolean combination of two atomic components: value description and neighbourhood description. A triple expression are inspired by regular expressions and restrict the expected neighbourhood of a node in RDF likewise DTDs and XML Schema in XML. A *neighbourhood description* defines the expected neighbourhood of a node and is given by a triple expression. While a *value description* is a set that declares the admissible values for a node: IRIs, literals and blank nodes.

To illustrate this, we consider the shape schema in Fig. 2 comprising one shape with shape label `ex:CustomerShape`, where `ex:CustomerShape` $\rightarrow$ `foaf:firstName xsd:string 1 ; foaf:lastName xsd:string 1, 3` is the shape expression. This shape expression contains a composed triple expression (two triple expressions separated by ';') constraining the relations `foaf:firstName` and `foaf:lastName` to be literals with a given cardinality.

ShEx defines cardinality bounds using regular expression conventions for cardinalities other than the default of "exactly one". The followings are the regular expressions considered in the ShEx Primer:

(a) '+': one or more

(b) '*': zero or more

(c) '?': zero or one

(d) '{m}': exactly m

(e) '{m,n}': at least m, no more than n.

It is easy to see that the last expression is the most generic and can be used to describe all the other expressions of cardinality. We then use the notation $\{min, max\}$ to build the ShEx shape expressions that serialise the cardinality bound of the relations.

For a given entity type τ we consider a set of cardinality bounds $\{\varphi_i\}_{i=1}^n$, where $\varphi_i = card(p_i, \tau) = (min.p_i, max.p_i)$. For every $i = 1, \ldots, n$, we use φ_i to build a ShEx schema using a template as the one in Fig. 6. Note that in the ShEx schema template, instead specifying a datatype we use a "." symbol that follows the relations. This symbol is interpreted by ShEx parsers and validators as a wildcard that could be replaced for any datatype (e.g., IRI, Literal, `xsd:string`, `xsd:integer`).

```
1  ex:myNewType {
2    p1 . {min.p1, max.p1} ;
3    p2 . {min.p2, max.p2} ;
4    ...
5    pn . {min.pn, max.pn}
6  }
```

Fig. 6. ShEx schema template for a set of cardinality bounds.

7 Evaluation

Next, we evaluate the application of our mining algorithm in its two variants against several real-world and synthetic KBs. These experiments aim to explore the usability of the mined cardinality bounds. The datasets and extracted ShEx shape schemas used in the paper are available publicly at http://emunoz.org/ kb-cardinality/, hoping it could serve other researchers.

In our experiments, we distinguish between real-world and synthetic datasets. Commonly, real-world datasets are more heterogeneous in nature and not all relations appear for a given entity type. However, synthetic datasets are usually generated automatically by programs that randomly create instance data from an input ontology, thus, the resulting instance data are more homogeneous. We hypothesise that these differences have twofold implications:

(a) synthetic KBs are usually more complete and consistent than real-world ones; and
(b) data shapes of incomplete entity types tend to have default bound values, i.e., cardinality bounds are usually 0 and ∞.

To evaluate these hypotheses, we perform two experiments:

(Exp-1) in which we use the cardinality bounds to assess completeness and consistency of entity types in the KBs; and
(Exp-2) in which we use the cardinality bounds to build ShEx shapes schema for entity types in the KBs.

7.1   Settings

Datasets. We used seven datasets with different characteristics such as number of triples and sameAs-axioms. They are diverse in domain of knowledge, features, and represent both real-world and synthetic data. We present their characteristics in Table 2 and describe them as follows:

Table 2. Datasets characteristics.

DATASET	№ TRIPLES	№ TYPES	№ PROP.	№ SAMEAS
LinkedMDB	3,579,532	41	148	92,589
OpenCyc	2,413,894	7,613	165	360,014
UOBM	2,217,286	40	29	0
British National Library	210,820	24	45	14,761
Mondial	186,534	27	60	0
New York Times People	103,496	1	20	14,884
SWDF	101,321	62	132	759

– LinkedMDB[11] is an open repository that describes movies, actors, directors, and so forth from the IMDB database.

---

[11] http://data.linkedmdb.org/.

- OpenCyc[12] is a large general KB released in 2012 that contains hundreds of thousands of terms in the domain of human knowledge covering places, organisations, business-related terms and people among others.
- UOBM[13] is a synthetic dataset that extends the Lehigh University Benchmark (LUMB), a university domain ontology, that contains information about faculties and students.
- British National Library[14] (BNL) is a dataset published by the National Library of the UK (second largest library in the world) about books and serials.
- Mondial[15] is a database compiled from geographical Web data sources such as CIA World Factbook, and Wikipedia.
- New York Times People[16] is a compilation of the most authoritative people mentioned in news of the New York Times newspaper since 2009.
- SWDF[17] is a small dataset containing information related to several semantic web related conferences and workshops.

Test Settings. We implemented our cardinality mining Algorithm 1 using Python 3.5 and Apache Spark 2.1.0. We use an Intel Core i7 4.0 GHz machine with 32 GB of RAM running Linux kernel 3.2 to run experiments on different KBs. Although Spark can run on multiple machines, we only tested it on a single machine using multiple parallel processes—one per core using 8 cores in total.

7.2   Experiment 1: Completeness and Consistency

In our first experiment, we measure the performance of both implementations—SPARQL and Spark—for Algorithm 1, and apply the mined cardinality bounds to assess the completeness and consistency of the source KBs. We then analyse qualitatively the use of the identified cardinality bounds to assess two crucial notions of KBs and databases, namely completeness and consistency.

Quantitative Evaluation. Intuitively, based on the scalability of Spark, one can foresee that the parallelised variant of our algorithm (cf. Fig. 5) should outperform the other using SPARQL. To test this we ran both implementations over two selected datasets, the British National Library and Mondial datasets, where only the former contains owl : sameAs axioms. The times correspond only to the extraction of the cardinality bounds and do not include the loading of data in memory or in the RDF store (triplestore).

For the BNL dataset, we fix the entity type to $\tau = $ Book, which co-occurs with 7 relations. We ran the code 10 times and obtained an average runtime of 253.908 ± 0.351 s for the SPARQL implementation, and 15.634 ± 0.118 s for the

---

[12] http://www.cyc.com/platform/opencyc.

[13] https://www.cs.ox.ac.uk/isg/tools/UOBMGenerator/.

[14] http://www.bl.uk/bibliographic/download.html.

[15] http://www.dbis.informatik.uni-goettingen.de/Mondial/.

[16] https://datahub.io/dataset/nytimes-linked-open-data.

[17] http://data.semanticweb.org/.

Spark one. This shows that the Spark implementation is 16x faster than the SPARQL implementation while performing the same task on the BNL dataset.

We repeat the same experiment over the Mondial dataset fixing the entity type τ = River, which co-occurs with 8 relations. We ran again the code 10 times and obtained an average runtime of 117.739 ± 0.651 s for the SPARQL implementation, and 2.948 ± 0.087 s for the Spark one. This shows that the Spark implementation is 40x faster than using SPARQL, while performing the same task with the Mondial dataset. The differences in the factors, 40x with the Mondial dataset and 16x with the BNL dataset, are due to the lower number of instances and the absence of sameAs-axioms in the data. These results show that the runtimes of the algorithm in either implementation are still small for two different but relatively small datasets. When dealing with much larger datasets such as DBpedia with 9.5 billion triples in its 2016-04 release, we can expect the runtimes to increase with the number of triples. Large datasets such as DBpedia grow in terms of entity types, triples per type, but also in terms of sameAs-axioms, where it will be interesting to test the performance of both implementations of Algorithm 1. Moreover, we believe that our Spark version will require several machines and run in its distributed manner. We leave such evaluations and other further optimisation as future work.

Finally, our experiments also show that the outlier detection method (i.e., box plot) does not add a significant overhead to the whole process and scales well (with the number of relations) for different data sizes.

Table 3. Evaluation of completeness and consistency per dataset: one type and five random properties per type.

ENTITY TYPE	№ sA-CLIQUES	№ TRIPLES BEFO./AFTER	COMPLET. RATIO	CONSIST. RATIO
Actor	92589	3579532/3536905	4/5	5/5
Fashion Model	118	1060/928	2/5	5/5
Research Assistant	0	135197/135197	4/5	5/5
Book	4515	97101/83556	2/5	3/5
Country	0	21766/21766	1/5	4/5
Concept	4979	58685/48780	2/5	5/5
InProceedings	759	101321/101302	0/5	5/5

Qualitative Evaluation. After showing that the mining of cardinality bounds is efficient in time, and to show the benefits of studying cardinality constraints derived from automatically discovered bounds in KBs, we bring to the fore their use on the assessment of data quality. Specifically, we evaluate each entity type in the dimensions of completeness and consistency from a common sense point of view. Because the consideration of cardinality bounds is application-dependent, here we try to abstract (without loss of generality) from individual use cases.

The cardinalities presented herein are considered robust bounds assessed to be a constraint by a knowledge engineer.

The characteristics of the studied datasets range between 1 up to 7,613 types and 20 up to 165 relations. To keep our study manageable, we selected randomly one entity type per dataset (7 in total) and five relations per type (35 in total). For each type, we show (cf. Table 3) the number of sameAs-cliques generated, and the number of triples before and after the rewriting process.

We consider that a relation p in the context of a type τ is *complete* given a cardinality constraint if every entity s of type τ has the "right number" of triples (s, p, o); and *incomplete* otherwise. For example, a constraint might be that all books must have at least one relation `title`, but the same is not true for relation `comment`. Also, we consider that a relation p in the context of a type τ is *consistent* if the triples with predicate p and subject s (of type τ) comply with the cardinality bounds; and *inconsistent* otherwise. For example, a constraint might be that all books must have always one `title`; however, we found five books which violate this constraint having two titles. Based on a set of verified discovered robust bounds, in Table 3 we show the ratios of completeness and consistency found in the five relations per type. For example, 2/5 completeness ratio in the entity type `Book` indicates that 2 out of 5 relations presented complete data, and the rest was incomplete. We did the same to measure consistency.

In general, we noticed a strong consistency and higher completeness on synthetic and curated datasets, where it is normal to define an ontology in which all instances are satisfied.

To further study the distribution of outliers in the cardinality bounds of each relation, we selected three entity types, namely, `foaf:Person`, `c4dm:Event` and `swrc:InProceedings` from the SWDF dataset. We plot the corresponding box plots for all relations in these entity types in Figs. 7, 8 and 9, respectively. These box plots provide us with the following insights for each entity type:

(1) `foaf:Person`: In this entity type we can see that the box plots are quite flat (i.e. they have small width). This suggests that overall the relation cardinalities have a high level of uniformity (agreement among them). Their values are centred around 1, thus, the mean cardinality is usually one. However, several probable outliers (denoted by black dots) are present in most of the relations. For instance, the relation `swc:holdsRole` is the one with more of them, going up to 22. `swc:holdsRole` records the roles of a person (e.g., PC member, chair, speaker, keynote) in a given event (e.g., conference, workshop), and shows that there is a huge variation of the cardinalities for different entities. Some people repeat more as organisers of events than others. A similar behaviour can be seen in the relation `foaf:made` used to state when a person is the author of a paper or a proceeding editor. Regarding the consistency of data, we can mention that two similar ontologies `swc` and `swrc` are used in entity instances of this type; however, equivalent relations in these ontologies do not match in the instance data. For instance, this can be seen in the `swrc:affiliation` and `swc:affiliation` relations, where

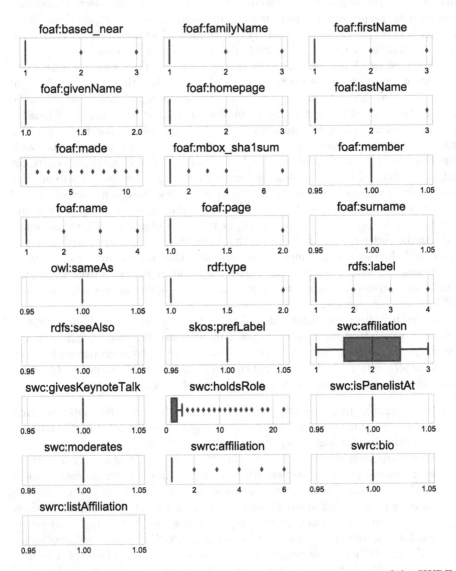

Fig. 7. Box plot figures for each relation in the entity type `foaf:Person` of the SWDF KB.

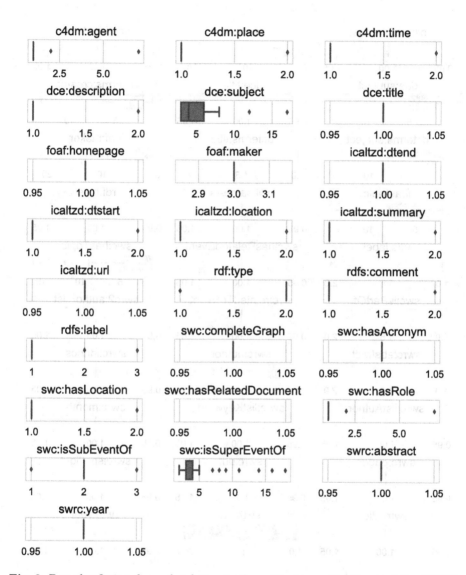

Fig. 8. Box plot figures for each relation in the entity type c4dm:Event of the SWDF KB.

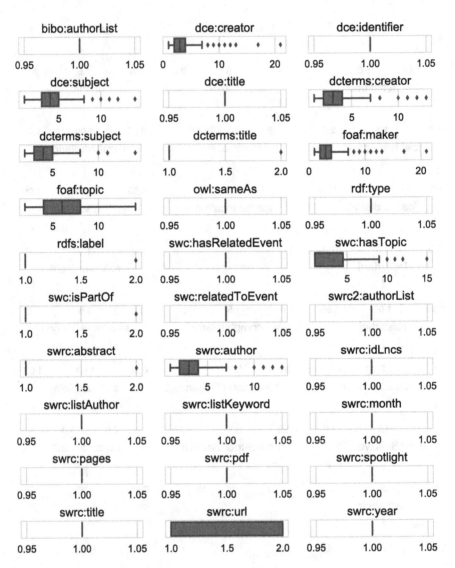

Fig. 9. Box plot figures for each relation in the entity type `swrc:InProceedings` of the SWDF KB.

the latter contains more variability than the former, but the former has more outliers.

(2) c4dm:Event: In this entity type we can also see quite flat box plots, mostly around the value 1, and with little outliers. Differences can be seen in the relations swc:isSuperEventOf (used to state sub-events of a conference or workshop, such as coffee/lunch breaks or talks) and dce:subject (used to state the topics of an event). The box plot of these two relations is larger in width meaning that the cardinalities are quite different but still within some low boundaries. Moreover, two other interesting cases are shown in the rdf:type and swc:isSubEventOf box plots, which present some outliers below the box plot. This is because the entities of this type have two values (e.g., swc:TalkEvent, c4dm:Event) for these relations in average, but few cases have only one.

(3) swrc:InProceedings: This entity type has more relations with wider box plots compared to the previous two and it also contains more outliers, indicating that cardinalities have high variability among entities. Some insights that we can collect by looking at the box plots are: (a) more than one relation coming from different ontologies is used to express the same attribute; and (b) box plots for relations that express the same attribute do not match in most cases. For instance, to illustrate (a) we have the case of the relations dce:creator, dcterms:creator, swrc:author, and foaf:maker that are used to represent the authors of a paper. However, some sets of entities contain only a subset of these relations. We believe that this depends on the metadata provided (when provided) by the conferences and workshops, which is not always the same. Likewise, the relations dce:subject and dcterms:subject show discordance in their data across entities.

The information collected from the box plots and cardinality bounds could be used to complete missing relations or spot inconsistencies in the data. Furthermore, we argue that by detecting cardinality inconsistencies and incompleteness we can determine structural problems at instance level. For example, if the KB is being automatically generated from a textual source, e.g., PubMed articles, structural problems such as missing labels could indicate issues in the pipeline generating the KB. In turn, this can be used to guide repair methods and move towards better quality of KBs.

```
1  card(mondial:name, mondial:Volcano)=(1,1)
2  card(mondial:elevation, mondial:Volcano)=(1,1)
3  card(mondial:longitude, mondial:Volcano)=(1,1)
4  card(mondial:latitude, mondial:Volcano)=(1,1)
5  card(mondial:locatedIn, mondial:Volcano)=(1,3)
6  card(mondial:lastEruption, mondial:Volcano)=(0,1)
```

Fig. 10. Cardinality bounds subset for entity type mondial:Volcano.

7.3 Experiment 2: Data Shape Inference

In our second experiment, we look at the mapping of the mined relation cardinality bounds to ShEx triple expressions that are used to build ShEx schemas that expose the shape of entity types. Such schemas could be used in tasks such as query formulation and optimisation, visualisations, and so forth [37].

To illustrate the benefits of having explicit cardinality bounds, we select the Mondial knowledge base and the entity type Volcano, and show in Fig. 10 a subset of the cardinality bounds extracted for this entity type. At a first glance, the extracted patterns correspond to what a knowledge engineer could expect: all volcanoes have a name and latitude/longitude coordinates; not all have information about their last eruption; and they have 1 to 3 locations for those that are in the intersection of different countries. Nevertheless, notice that even when we obtain upper bounds for cardinalities (e.g., relation mondial:locatedIn with upper bound of 3), in reality upper bounds could be higher and should not be considered as a constraint for validation until a user assessment. We found a similar situation with relation mondial:hasCity in the type mondial:Country, where the robust cardinality identified was $card(\{mondial:hasCity\}, mondial:Country) = (1,31)$. However, based on that upper bound, there are 25 identified "outliers", whose cardinalities are indeed abnormal in the distribution, but they are not wrong. In the instance data, we can find country instances like China with 306 cities, USA with 250 cities, Brazil with 210 cities, Russia with 171 cities, and India with 99 cities, which are outside of the range considered as robust but they are still valid values. With that information at hand, data users and knowledge engineers could determine whether a given relation cardinality bound should be promoted to become a constraint (i.e., verified discovered robust bounds) or not.

To demonstrate the serialisation of relation cardinality bounds as ShEx schemas, we select three entity types from the SWDF dataset and generate their corresponding ShEx schemas following the mapping described in Sect. 6. Figure 11(a) shows a sample of triples in Turtle format for the entity types, namely, foaf:Person, c4dm:Event and swrc:InProceedings. For each one of these entity types, we generate a different colour-coded ShEx schema using the relation cardinality bounds extracted from the instance data as shown in Fig. 11(b). We notice that the instance data perfectly matches the cardinality constraints stated in the schemas. Indeed, it is easy to see from the data shapes what properties correspond to each entity type, thus, allowing data users to gain insights on how to query and use the data.

Furthermore, from the shapes in Fig. 11(b) we can extract more relevant information. For instance, it is very infrequent to find lower bounds different from zero showing that most properties are not required all the time in the tested entity types. On the other hand, some properties that one could think as functional for the entity types, e.g., foaf:familyName for foaf:Person, dc:title for c4dm:Event and bibo:authorList for swrc:InProceedings are not. This could indicate some serious incompleteness in the generation of KBs, and pose a big challenge while querying the data. Tools and systems that consume the data

```
sw:person/aidan-hogan  swc:holdsRole      sw:conference/ekaw/2012/programme-committee-member,
                                          sw:conference/eswc/2010/social_web/pcmember,
                                          sw:conference/eswc/2012/research-track-committee-member,
                                          sw:conference/eswc/2013/eswc-2013-track-committee-member;
                       swcr:affiliation   sw:organization/deri-nui-galway;
                       swcr:listAffiliation "DERI, NUI Galway" ;
                       foaf:based_near    dbr:Ireland ;
                       foaf:made          sw:conference/ekaw/2012/paper/research/71,
                                          sw:conference/workshop/ldow/2012/paper/42 ;
                       ...                ...
                       rdf:type           foaf:Person .

sw:workshop/ldow/2012  swc:hasAcronym        "LDOW2012" ;
                       swc:hasLocation       sw:conference/www/2012/location/54 ;
                       swc:hasRelatedDocument sw:workshop/ldow/2012/proceedings ;
                       swc:hasRole           sw:workshop/ldow/2012/programme-chair,
                                             sw:workshop/ldow/2012/programme-committee-member ;
                       dc:subject            "Evolution" ,
                                             "Linked Data" ,
                                             "Web of Data" ;
                       dc:title              "Linked Data on the Web" ;
                       c4dm:place            sw:workshop/conference/www/2012/location/54 ;
                       swrc:year             "2012" ;
                       ...                   ...
                       rdf:type              c4dm:Event .

sw:conference/www/2012/paper/809
                       swc:isPartOf       sw:conference/www/2012/proceedings ;
                       dc:creator         sw:person/jorge-a-perez ,
                                          sw:person/marcelo-arenas ,
                                          sw:person/sebastian-conca ;
                       dc:subject         "bag semantics" ,
                                          "combined complexity" ,
                                          "property paths" ;
                       dc:title           "Counting beyond a Yotabyte, or how SPARQL 1.1 Property Paths
                                          will prevent adoption of the standard" ;
                       swrc:month         "April" ;
                       swrc:year          "2012" ;
                       foaf:maker         sw:person/jorge-a-perez ,
                       ...                ...
                       rdf:type           swrc:InProceedings .
```

(a) Sample of triples in turtle format

```
foaf:Person {                          c4dm:Event {                           swrc:InProceedings {
  swc:affiliation . {0, 3} ;             swc:completeGraph . {0, 1} ;           swc:hasRelatedEvent . {0, 1} ;
  swc:givesKeynoteTalk . {0, 1} ;        swc:hasAcronym . {0, 1} ;              swc:hasTopic . {0, 15} ;
  swc:holdsRole . {0, 22} ;              swc:hasLocation . {0, 2} ;             swc:isPartOf . {0, 2} ;
  swc:isPanelistAt . {0, 1} ;            swc:hasRelatedDocument . {0, 1} ;      swc:relatedToEvent . {0, 1} ;
  swc:moderates . {0, 1} ;               swc:hasRole . {0, 7} ;                 dc:creator . {0, 21} ;
  swrc:affiliation . {0, 6} ;            swc:isSubEventOf . {0, 3} ;            dc:identifier . {0, 1} ;
  swrc:bio . {0, 1} ;                    swc:isSuperEventOf . {0, 18} ;         dc:subject . {0, 14} ;
  swrc:listAffiliation . {0, 1} ;        c4dm:agent . {0, 7} ;                  dc:title . {0, 1} ;
  rdfs:label . {0, 4} ;                  c4dm:place . {0, 2} ;                  dct:creator . {0, 13} ;
  rdfs:seeAlso . {0, 1} ;                c4dm:time . {0, 2} ;                   dct:subject . {0, 14} ;
  owl:sameAs . {0, 1} ;                  dc:description . {0, 2} ;              dct:title . {0, 2} ;
  skos:prefLabel . {0, 1} ;              dc:subject . {0, 17} ;                 bibo:authorList . {0, 1} ;
  foaf:based_near . {0, 3} ;             dc:title . {0, 1} ;                    swrc:abstract . {0, 2} ;
  foaf:familyName . {0, 3}               swrc:abstract . {0, 1} ;               swrc:author . {0, 13} ;
  foaf:firstName . {0, 3} ;              swrc:year . {0, 1} ;                   swrc:idLncs . {0, 1} ;
  foaf:givesName . {0, 2} ;              rdfs:comment . {0, 2} ;                swrc:listAuthor . {0, 1} ;
  foaf:homepage . {0, 3} ;               rdfs:label . {1, 3} ;                  swrc:listKeyword . {0, 1} ;
  foaf:lastName . {0, 3} ;               ical:dtend . {0, 1} ;                  swrc:month . {0, 1} ;
  foaf:made . {0, 11} ;                  ical:dtstart . {0, 2} ;                swrc:pages . {0, 1} ;
  foaf:mbox_sha1sum . {0, 7} ;           ical:location . {0, 2} ;               swrc:pdf . {0, 1} ;
  foaf:member . {0, 1} ;                 ical:summary . {0, 2} ;                swrc:spotlight . {0, 1} ;
  foaf:name . {1, 4} ;                   ical:url . {0, 1} ;                    swrc:title . {0, 1} ;
  foaf:page . {0, 2} ;                   foaf:homepage . {0, 1} ;               swrc:url . {0, 2} ;
  foaf:surname . {0, 1}                  foaf:maker . {0, 3}                    swrc:year . {0, 1} ;
}                                      }                                       swrce:authorList . {0, 1} ;
                                                                               rdfs:label . {0, 2} ;
                                                                               owl:sameAs . {0, 1} ;
                                                                               foaf:maker . {0, 21} ;
                                                                               foaf:topic . {0, 14}
                                                                             }
```

(b) Extracted data shapes for three entity types

Fig. 11. Data shape extraction from the Semantic Web Dog Food dataset.

should be aware of that incompleteness and generate adequate queries. Now this is possible with these data shapes.

8 Conclusion

In this paper, we have studied the notion of relation cardinality bounds in knowledge bases that have been hitherto mostly neglected. We propose a notion of cardinality as a solution to partially cope with the lack of explicit structure of KBs on the Web. In summary, we have shown some of the practical benefits that relation cardinality bounds provide: (1) They represent characteristics that the KB naturally exhibits, independently from the use of ontologies in the data. (2) They allow us to assess two important indicators of data quality: completeness and consistency. (3) They unlock a good KB data design by making explicit the structure as data shapes. Since cardinalities are not usually declared in KBs, to take advantage of these benefits we proposed a mining algorithm that takes a KB and returns accurate and robust relation cardinality bounds. Although the cardinality information extracted in this way should be considered as *soft* constraints, they have to be uplifted to *hard* constraints before they can be used to validate KBs data. We presented two implementations of our approach, namely SPARQL- and Spark-based, and evaluated them against seven different datasets including real-world and synthetic data. Our results show that cardinality bounds can be mined efficiently, and are useful to understand the structure of data at a glance, and also to analyse the completeness and consistency of data.

Limitations and Future Work. Although our results are promising, there are some limitations to our work and possible open challenges that we highlight next. During the normalisation step, we consider only equality between entities; however, in RDF equality could also be defined between relations, e.g., relations `ex:hasPart` is equivalent to the inverse of the relation `ex:partOf`. OWL has the `owl:equivalentProperty` construct used to state that two relations have the same extension or set of target nodes meaning that they are equivalent. Such axioms could also be considered during the rewriting process to capture even better the semantics of data. Regarding our experiments, we have not reported on the runtime of our Spark-based implementation over several machines. This is a requirement to evaluate large datasets such as DBpedia with 9.5 billion triples. Theoretically, an even better performance could be obtained by running the software in a cluster setup. We leave this as potential future work, where other parallelisation techniques could also be applied. We have also limited our study to qualified cardinality bounds and leave unqualified bounds for future work. Unqualified bounds could help in situations where a relation's semantic is unchangeable by its context, i.e., appearance in a given entity type. For instance, the relations `skos:prefLabel` and `rdfs:label` are used in the same way while describing instances of most entity types.

In practical terms, cardinalities have been used to enable both logical and physical query optimisation in databases, and we believe that they could do the same for Semantic Web KBs [38]. In [25] the authors present some thoughts about schema-based optimisation techniques for joins using cardinalities defined in OWL. They indicate how cardinalities can be used to guide the optimiser into selecting more efficient orders for join operations.

Finally, we found recent approaches that use cardinality of properties to inform rule mining algorithms in KBs [10,39]. Nonetheless, such approaches do not consider neither sameAs-axioms nor statistical outliers. Finally, we believe that cardinality bounds could be an important contribution to approaches that perform data mining and knowledge discovery over KBs.

Acknowledgements. This work has been supported by TOMOE project funded by Fujitsu Laboratories Ltd., Japan and Insight Centre for Data Analytics at National University of Ireland Galway, Ireland.

References

1. Abiteboul, S., Hull, R., Vianu, V.: Foundations of Databases. Addison-Wesley, Boston (1995)
2. Arenas, M., Conca, S., Pérez, J.: Counting beyond a yottabyte, or how SPARQL 1.1 property paths will prevent adoption of the standard. In: WWW, pp. 629–638. ACM (2012)
3. Arenas, M., Gutierrez, C., Pérez, J.: Foundations of RDF databases. In: Tessaris, S., et al. (eds.) Reasoning Web 2009. LNCS, vol. 5689, pp. 158–204. Springer, Heidelberg (2009). https://doi.org/10.1007/978-3-642-03754-2_4
4. Boneva, I., Labra Gayo, J.E., Prud'hommeaux, E.G.: Semantics and validation of shapes schemas for RDF. In: d'Amato, C., et al. (eds.) ISWC 2017. LNCS, vol. 10587, pp. 104–120. Springer, Cham (2017). https://doi.org/10.1007/978-3-319-68288-4_7
5. Bosch, T., Eckert, K.: Guidance, please! towards a framework for RDF-based constraint languages. In: Proceedings of the International Conference on Dublin Core and Metadata Applications (2015)
6. Chandola, V., Banerjee, A., Kumar, V.: Anomaly detection: a survey. ACM Comput. Surv. **41**(3), 15:1–15:58 (2009)
7. Christodoulou, K., Paton, N.W., Fernandes, A.A.A.: Structure inference for linked data sources using clustering. Trans. Large-Scale Data Knowl.-Centered Syst. **19**, 1–25 (2015)
8. Ferrarotti, F., Hartmann, S., Link, S.: Efficiency frontiers of XML cardinality constraints. Data Knowl. Eng. **87**, 297–319 (2013)
9. Fleischhacker, D., Paulheim, H., Bryl, V., Völker, J., Bizer, C.: Detecting errors in numerical linked data using cross-checked outlier detection. In: Mika, P., et al. (eds.) ISWC 2014. LNCS, vol. 8796, pp. 357–372. Springer, Cham (2014). https://doi.org/10.1007/978-3-319-11964-9_23
10. Galárraga, L., Razniewski, S., Amarilli, A., Suchanek, F.M.: Predicting completeness in knowledge bases. In: WSDM, pp. 375–383. ACM (2017)
11. Glimm, B., Hogan, A., Krötzsch, M., Polleres, A.: OWL: yet to arrive on the web of data? In: LDOW. CEUR Workshop Proceedings, vol. 937. CEUR-WS.org (2012)
12. Hogan, A., Harth, A., Passant, A., Decker, S., Polleres, A.: Weaving the pedantic web. In: LDOW. CEUR Workshop Proceedings, vol. 628. CEUR-WS.org (2010)

13. Horrocks, I., Tessaris, S.: Querying the semantic web: a formal approach. In: Horrocks, I., Hendler, J. (eds.) ISWC 2002. LNCS, vol. 2342, pp. 177–191. Springer, Heidelberg (2002). https://doi.org/10.1007/3-540-48005-6_15

14. Kellou-Menouer, K., Kedad, Z.: Evaluating the gap between an RDF dataset and its schema. In: Jeusfeld, M.A., Karlapalem, K. (eds.) ER 2015. LNCS, vol. 9382, pp. 283–292. Springer, Cham (2015). https://doi.org/10.1007/978-3-319-25747-1_28

15. Kostylev, E.V., Reutter, J.L., Romero, M., Vrgoč, D.: SPARQL with property paths. In: Arenas, M., et al. (eds.) ISWC 2015. LNCS, vol. 9366, pp. 3–18. Springer, Cham (2015). https://doi.org/10.1007/978-3-319-25007-6_1

16. Lausen, G., Meier, M., Schmidt, M.: SPARQLing constraints for RDF. In: EDBT, ACM International Conference Proceeding Series, vol. 261, pp. 499–509. ACM (2008)

17. Liddle, S.W., Embley, D.W., Woodfield, S.N.: Cardinality constraints in semantic data models. Data Knowl. Eng. **11**(3), 235–270 (1993)

18. Motik, B., Horrocks, I., Sattler, U.: Bridging the gap between OWL and relational databases. J. Web Sem. **7**(2), 74–89 (2009)

19. Motik, B., Nenov, Y., Piro, R.E.F., Horrocks, I.: Handling Owl:sameAs via rewriting. In: AAAI, pp. 231–237. AAAI Press (2015)

20. Motik, B., Patel-Schneider, P.F., Parsia, B.: OWL 2 Web Ontology Language Structural Specification and Functional-Style Syntax, 2nd edn (2012). http://www.w3.org/TR/2012/REC-owl2-syntax-20121211/

21. Muñoz, E.: On learnability of constraints from RDF data. In: Sack, H., Blomqvist, E., d'Aquin, M., Ghidini, C., Ponzetto, S.P., Lange, C. (eds.) ESWC 2016. LNCS, vol. 9678, pp. 834–844. Springer, Cham (2016). https://doi.org/10.1007/978-3-319-34129-3_52

22. Muñoz, E., Nickles, M.: Mining cardinalities from knowledge bases. In: Benslimane, D., Damiani, E., Grosky, W.I., Hameurlain, A., Sheth, A., Wagner, R.R. (eds.) DEXA 2017. LNCS, vol. 10438, pp. 447–462. Springer, Cham (2017). https://doi.org/10.1007/978-3-319-64468-4_34

23. Neumann, T., Moerkotte, G.: Characteristic sets: accurate cardinality estimation for RDF queries with multiple joins. In: ICDE, pp. 984–994. IEEE Computer Society (2011)

24. Olivé, A.: Conceptual Modeling of Information Systems. Springer, Heidelberg (2007). https://doi.org/10.1007/978-3-540-39390-0

25. Papakonstantinou, V., Flouris, G., Fundulaki, I., Gubichev, A.: Some thoughts on OWL-empowered SPARQL query optimization. In: Sack, H., Rizzo, G., Steinmetz, N., Mladenić, D., Auer, S., Lange, C. (eds.) ESWC 2016. LNCS, vol. 9989, pp. 12–16. Springer, Cham (2016). https://doi.org/10.1007/978-3-319-47602-5_3

26. Paulheim, H.: Knowledge graph refinement: a survey of approaches and evaluation methods. Semant. Web **8**(3), 489–508 (2017)

27. Paulheim, H., Bizer, C.: Improving the quality of Linked Data using statistical distributions. Int. J. Semant. Web Inf. Syst. **10**(2), 63–86 (2014)

28. Pearson, R.K.: Mining imperfect data - dealing with contamination and incomplete records. SIAM (2005)

29. Polleres, A., Reutter, J.L., Kostylev, E.V.: Nested constructs vs. sub-selects in SPARQL. In: AMW. CEUR Workshop Proceedings, vol. 1644. CEUR-WS.org (2016)

30. Polleres, A., Scharffe, F., Schindlauer, R.: SPARQL++ for mapping between RDF vocabularies. In: Meersman, R., Tari, Z. (eds.) OTM 2007. LNCS, vol. 4803, pp. 878–896. Springer, Heidelberg (2007). https://doi.org/10.1007/978-3-540-76848-7_59

31. Prud'hommeaux, E., Gayo, J.E.L., Solbrig, H.R.: Shape expressions: an RDF validation and transformation language. In: SEMANTICS, pp. 32–40. ACM (2014)
32. Rivero, C.R., Hernández, I., Ruiz, D., Corchuelo, R.: Towards discovering ontological models from big RDF data. In: Castano, S., Vassiliadis, P., Lakshmanan, L.V., Lee, M.L. (eds.) ER 2012. LNCS, vol. 7518, pp. 131–140. Springer, Heidelberg (2012). https://doi.org/10.1007/978-3-642-33999-8_16
33. Rosner, B.: Percentage points for a generalized ESD many-outlier procedure. Technometrics 25(2), 165–172 (1983)
34. Russell, S.J., Norvig, P.: Artificial Intelligence - A Modern Approach, 3rd internat. edn. Pearson Education (2010)
35. Ryman, A.G., Hors, A.L., Speicher, S.: OSLC resource shape: a language for defining constraints on linked data. In: LDOW. CEUR Workshop Proceedings, vol. 996. CEUR-WS.org (2013)
36. Schenner, G., Bischof, S., Polleres, A., Steyskal, S.: Integrating distributed configurations with RDFS and SPARQL. In: Configuration Workshop. CEUR Workshop Proceedings, vol. 1220, pp. 9–15. CEUR-WS.org (2014)
37. Schmidt, M., Lausen, G.: Pleasantly consuming linked data with RDF data descriptions. In: COLD. CEUR Workshop Proceedings, vol. 1034. CEUR-WS.org (2013)
38. Schmidt, M., Meier, M., Lausen, G.: Foundations of SPARQL query optimization. In: ICDT, pp. 4–33. ACM International Conference Proceeding Series. ACM (2010)
39. Tanon, T.P., Stepanova, D., Razniewski, S., Mirza, P., Weikum, G.: Completeness-aware rule learning from knowledge graphs. In: d'Amato, C., et al. (eds.) ISWC 2017. LNCS, vol. 10587, pp. 507–525. Springer, Cham (2017). https://doi.org/10.1007/978-3-319-68288-4_30
40. Thalheim, B.: Fundamentals of cardinality constraints. In: Pernul, G., Tjoa, A.M. (eds.) ER 1992. LNCS, vol. 645, pp. 7–23. Springer, Heidelberg (1992). https://doi.org/10.1007/3-540-56023-8_3
41. Töpper, G., Knuth, M., Sack, H.: DBpedia ontology enrichment for inconsistency detection. In: I-SEMANTICS, pp. 33–40. ACM (2012)
42. Vandenbussche, P., Atemezing, G., Poveda-Villalón, M., Vatant, B.: Linked open vocabularies (LOV): a gateway to reusable semantic vocabularies on the web. Semant. Web 8(3), 437–452 (2017)
43. Völker, J., Niepert, M.: Statistical schema induction. In: Antoniou, G., et al. (eds.) ESWC 2011. LNCS, vol. 6643, pp. 124–138. Springer, Heidelberg (2011). https://doi.org/10.1007/978-3-642-21034-1_9
44. Wienand, D., Paulheim, H.: Detecting incorrect numerical data in DBpedia. In: Presutti, V., d'Amato, C., Gandon, F., d'Aquin, M., Staab, S., Tordai, A. (eds.) ESWC 2014. LNCS, vol. 8465, pp. 504–518. Springer, Cham (2014). https://doi.org/10.1007/978-3-319-07443-6_34

ETL Processes in the Era of Variety

Nabila Berkani[1]([✉]), Ladjel Bellatreche[2], and Laurent Guittet[2]

[1] Ecole nationale Supérieure d'Informatique, BP 68M,
16309 Oued-Smar, Alger, Algérie
n_berkani@esi.dz
[2] LIAS/ISAE-ENSMA – Poitiers University, Futuroscope, France
{bellatreche,guittet}@ensma.fr

Abstract. Nowadays, we are living in an open and connected world, where small, medium and large companies are looking for integrating data from various data sources to satisfy the requirements of new applications such as delivering real-time alerts and trigger automated actions, complex system failure detection, anomalies detection, etc. The process of getting these data from their sources to its home system in efficient and correct manner is known by data ingestion, usually refer to Extract, Transform, Load (ETL) widely studied in data warehouses. In the context of rapidly technology changing and the explosion of data sources, ETL processes have to consider two main issues: (a) the variety of data sources that spans traditional, XML, semantic, graph databases, etc. and (b) the variety of storage platforms, where the home system may have several stores (known by polystore), where one hosts a particular type of data. These issues directly impact the efficiency and the deployment flexibility of ETL. In this paper, we deal with these issues. Firstly, thanks to Model Driven Engineering, we make generic different types of data sources. This genericity allows overloading the ETL operators for each type of sources. This genericity is illustrated by considering three types of the most popular data sources: relational, semantic and graph databases. Secondly, we show the impact of genericity of operators in the ETL workflow, where a Web-service-driven approach for orchestrating the ETL flows is given. Thirdly, the extracted and merged data obtained by the ETL workflow are deployed according their favorite stores. Finally, our finding is validated through a proof of concept tool using the LUBM semantic database and Yago graph deployed in Oracle RDF Semantic Graph 12c.

1 Introduction

Competitiveness in development of countries, regions, universities and companies has become a critical challenge at the time of integration and globalization of growing economies. By searching on Google via the term competitiveness, for instance, 34,000,000 search results that focus on several aspects ranging from presidential elections to job creation or cutting, and so forth. This shows a great interest at now. As specifically regards companies, competitiveness is usually connected to value-add-chain productivity. Certainly, big data technologies mainly

© Springer-Verlag GmbH Germany, part of Springer Nature 2018
A. Hameurlain et al. (Eds.): TLDKS XXXIX, LNCS 11310, pp. 98–129, 2018.
https://doi.org/10.1007/978-3-662-58415-6_4

focus on the dimension value, but better inspecting typical activities of companies, we argue that value-add is emerging among the most relevant big data dimensions for companies and their assets. This evidence has been confirmed by the relevant efforts and resources that companies have invested recently in order to achieve this goal. These efforts specifically concern with recruitment, data analysis, development of services, material deployment machines, etc. From a data-analysis perspective, this means that companies follow the so-called CSAD chain, namely: Collecting data from local sources, Storing data, Analyzing data, and Discovering hidden knowledge from data.

Historically, the collection and storage activities pass through the data integration process [17]. This integration can be either materialized or virtual [6]. In the 1990's, the data warehouse (DW) and its surrounding services and tools have been launched to respond to the analytical activities. The DW is an example of materialized data integration system. It is defined as a subject-oriented, integrated, time-variant, non-volatile collection of data in support of management's decision-making process [12].

ETL (Extract, Transform, Load) activities play a crucial role in conducting the CSAD chain [2]. It represents the process of getting data out from sources and load it into a data warehouse. This data has also to be transformed to match the DW schema. ETL is totally performed outside the DW, in the staging area. To synchronize the data between the sources and the DW two approaches exist: complete rebuild which periodically re-build the DW from the sources (e.g. every night) and incremental update which periodic updates based on the changes in the sources are performed. ETL activities cooperate all together and generate ETL workflow [33]. Each activity is associated to one or several ETL operators [28] of the ETL algebra composed of 10 operators (Retrieve, Extract, Convert, Filter, Merge, Join, Union, Aggregate, Delete and Store). These operations can be divided into two main categories: (i) operators for sources (Extract, Retrieve, Merge, Union, Join, Filter). They associated to activities related to data sources. (ii) Operators for store (e.g., Store) that takes care of loading extracted and transformed data to the warehouse store. Each operator is then personalized according the type of data sources and the target DW ([33] for traditional sources, [5, 20] for semantic sources, etc.). A plenty of commercial and open sources ETL are available in the market. We can cite Oracle Warehouse Builder, SAP Data Service, Talent Studio for Data Integration, IBM Infosphere Warehouse Edition, etc. Each tool uses a different approach for modeling ETL activities [34]. To unify the different modeling approaches, several research efforts have been deployed to make generic the ETL workflow, standard languages and tools were used: UML [31], model driven architecture (MDA) [13], BPMN [1, 38]. Having this genericity facilitates the maintenance [34], the tractability [15] and deployment of ETL workflow [5].

The unification efforts have concerned only ETL workflow and some parts of activities. We claim that this unification has to cover the ETL environment that includes the following elements: activities usually associated to operators, workflow, the data sources, the DW schema and its store.

The unification of data sources allows dealing with their variety. This variety entails heterogeneity that may concert data formats, meaning, context, etc. In the first generations of DW, the involved data source types cover csv files, traditional databases, XML documents, semantic data, etc. In the context of rapidly technology changing, several new types of databases appear (e.g., Graph databases, NoSQL, Time Series, Knowledge bases (KB), linked open data and consequently they became candidate for the data integration. This phenomenon increases the degree of the variety of sources [10].

The variety does not only impact data sources, but also the storage of the DW, where multi-stores (polystore) are well-adapted to achieve high performance of data accesses (no one size fits all [29]). In the past decade, the database community has seen an explosion data management systems, where each targeted for a well-defined vertical market [11]. Instead of deploying in one store as in traditional DW applications, polystore deployment has become a reality. The variety is present at the source level and the deployment level as shown in Fig. 1. More concretely, we passed from $(n - 1)$ scenario, where n heterogeneous sources are integrated into DW deployed on one store to $(n - m)$ scenario, where the DW may be deployed in several stores, where each one may store a specific type of data [16].

Based on this discussion, we realize that to get added-value of companies, we need to deal with the variety. As result, we can say that the value and variety are strongly dependent – to get added-value, we need to deal with the high variety.

To contribute to the generalization of whole ETL environment, we propose to make generic all schemes of data sources and the ETL operators. This passes through the usage of Meta-Driven Engineering (MDE) and then overload the ETL operators to deal with specific type of source. Inspired from the Meta-Object Facility (MOF). The MOF describes a generic framework in which the abstract syntax of modelling languages can be defined [22]. It has a hierarchical structure composed of four layers of meta-data corresponding to the different levels of abstraction: the instance layer (M0), the model layer (M1), the meta-model layer (M2) and the meta-meta-model layer (M3). Each layer defines a level to ensure the consistency and the correctness of the instance model syntax and semantics at each level of abstraction.

This generic model can be easily exploited by ETL operators, where each one is overload. An operator overloading (as in C++ language) allows a programmer to define the behavior of an operator applied to objects of a certain class the same way methods are defined [8]. In our context, each ETL operator will be overload to deal with the diversity of each type of sources.

To offer the designers the possibility to deploy their DW on a polystore, we exploit the Store operator of ETL. It can be associated with a Service Web that orchestrates the ETL workflow and distributes the data over the stores according their storage formats (Fig. 2).

In this paper that represents an extension of [4], we detail our generic model using MDE. We give examples of its instantiation from three types of data sources: relational, semantic and graph databases. The ETL operators are then

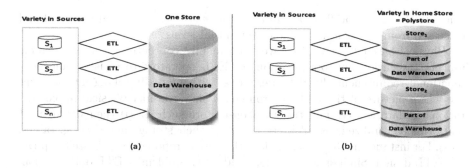

Fig. 1. Variety at sources and store levels

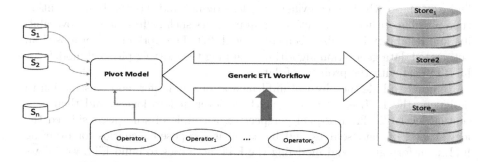

Fig. 2. Generalized unification of ETL environment

overloaded for these types. Thanks to the Store operator, the multi-store deployment is guaranteed. Our proposal is implemented and experimented.

The rest of this paper is organized as follows. We give an overview on the evolution of the ETL in Sect. 2. In Sect. 3, we present a meta model of ETL workflow that allows a generic modeling of the activities of the ETL environment. In Sect. 4, we give a formalization of three main classes of databases (relational, semantic and graph) and a motivating example. In Sect. 5, a generalization of ETL elements are given by the means of MDE techniques and the process to overload the ETL operators. In Sect. 6, the deployment methodology of a data warehouse on multi-store system is developed. In Sect. 7, a case study is proposed and various experiments are presented. Section 8 concludes the paper.

2   Related Work

In this section, we overview the efforts of unifying the elements of the ETL environment to deal with the variety of data sources and polystore deployment.

Unification of the ETL Environment. As any process, designers of ETL have to model the ETL environment and then execute its elements. An important number of studies were mainly concentrated on ETL activities modeling to generate the ETL workflow. Historically, these studies have been proposed and discussed from physical, logical and conceptual levels.

Physical-based ETL works have been mainly concentrated on algorithmic aspects (execution and optimization) of ETL operators to satisfy some non-functional requirements fixed by designers, without really taking into account the whole picture of the ETL environment. They consider physical implementations of data sources and the warehouse store such as: (i) their deployment platforms (centralized, parallel, etc.) and (ii) their storage models (e.g. tables, files). For instance, in [32], a set of algorithms was proposed to optimize the physical ETL design. Simitsis and al. [25] handle the problem of ETL optimization. It mainly focuses on the order of activities in the workflow. Given an original ETL scenario provided by the warehouse designer, the goal of the paper is to find a scenario which is equivalent to the original and has the best execution time possible. Other non-functional requirements such as freshness, recoverability, and reliability have also been considered [27]. The work of [19] proposed an automated data generation algorithm assuming the existing physical models for ETL to deal with the problem of data growing.

In order to hide the physical implementations and offer more abstraction to designers, the ETL environment has been leveraged to logical and then conceptual levels [2]. From logical perspective, a couple of studies considered the logical level of data sources and based on it, proposed scenarios for ordering ETL transformations and modeling the ETL activities to build ETL workflows by the means of ad hoc templates and taxonomies. They considered the sources at the M1 level of the MOF. In this perspective, [37] proposed an ETL workflow modeled as a graph, where its nodes represent activities, record-sets, attributes, and its edges describe the relationships between nodes that define ETL transformations. In [35], a formal ETL logical model is given using L_{DL} [24] as a formal language for expressing the operational semantics of ETL activities.

To offer more abstraction, the ETL environment has been modeled as conceptual level. The work in this direction can be classified at M2 level of the MOF as long as they attempt to propose meta-models for ETL modeling based on: ad-hoc formalisms [36], standard languages using UML [31], model driven architecture (MDA) [13] and BPMN [1,38]. Mapping modeling [7,18] also have been proposed.

Due to the similarity between conceptual models and ontologies, some research efforts have considered ontologies as external resources to facilitate and automate the conceptual design of ETL process. [28] automated the ETL process by constructing an OWL ontology linking schemes of semi-structured and structured (relational) sources to a target data warehouse ($\mathcal{DW}$) schema. Other studies like [20] consider data source provided by the semantic Web and annotated by OWL ontologies. However, the ETL process in this work is dependent on the storage model used for instances which is the triples. [28] has defined an ETL algebra with 10 generic operators in the context of semantic sources.

Based on this brief overview, we figure out the effort that has been deployed by the research community in generalizing the ETL environment. The ultimate goal is to uniform the modeling of ETL operators and activities. Our observation is that existing works have evolved according to the different layers of the MOF model. In [37], a generic model of ETL activities that plays the role of a pivot model has been proposed. Mappings between different levels: conceptual $\longrightarrow$ logical $\longrightarrow$ physical are not deeply studied.

Ultimately, all these works aspire to propose a generic and archetype model combining ETL activities that involves ETL operators during the execution of the ETL workflow. However, the main drawbacks of these approaches are: (i) the lack of a uniform algebra and signature to describe the ETL operators and (ii) a reference way to represent ETL activities that allow the efficient orchestration of the ETL operators. In addition, existing works deal only with traditional types of sources (relational and XML schemes) and they make an implicit assumption that the data warehouse is deployed on one system usually relational. Besides, no work attempts to propose a pivot and uniform model to cover a large variety of ETL activities with complicated semantics and provide the framework for the optimization of ETL flows. The need being to make generic the signature of the ETL operators orchestrated by ETL activities. In our work, we meet this problem by raising the level of abstraction at M2, pivot meta-model, of MOF hierarchy allowing dealing with a large variety of data sources.

A comparison of main important studies and our proposal is given in Table 1 based on five main criteria: (1) The type of data sources involved: Relational (R), Semantic (S) or Graph (G); (2) the objective of the proposal: Modeling (M), non-functional requirement satisfaction (NfRSat), (3) the design levels of ETL: physical (P), logical (L) or conceptual (C); (4) the type of deployment store: one store (1-Store) and polystore (PS) and (5) the parallel with the MOF levels: M0–M3.

Table 1. Comparison of ETL process works

Work	Criteria				
	Data sources	Goal	Design level	Deployment	MOF level
Vassiliadis et al. [35,37]	Relational	Modeling	L/P	1-store	M1
Trujillo et al. [31]	Relational	M	C	1-store	M2
Alkis et al. [25]	Relational	M/NfRSat	L/P	1-store	M1
Tziovara et al. [32]	Relational	NfRSat	P	1-store	M0
Skoutas et al. [28]	Semantic	M/NfRSat	C	1-store	M2
Mazon et al. [13]	Relational	Modeling	C	1-store	M2
Wilkinson et al. [1,38]	Relational	Modeling	C	1-store	M2
Simitsis et al. [27]	Relational	NfRSat	P	1-store	M0
Nebot et al. [20]	Semantic	M/NfRSat	C/L/P	1-store	M2
Our proposal	R/S/G	M/NfRSat	C/L/P	PS	M2 pivot

2.1   Multi-store Systems

Polystore systems provide integrated access to data stores such as RDBMS, NoSQL or HDFS. In this section, we briefly overview a set of representative systems that exists in the literature and their connection with loading operation of ETL.

The BigIntegrator [39] is a system that combines cloud-based data stores with relational databases. It provides a query processing mechanism based on developed plugins (absorbers and finalizers) to allow extensions of new kind data source. It uses the LaV (Local as View) approach for defining the global schema of the Bigtable and relational data sources as flat relational tables. Each store contains several collections represented as a source table of the form *table-name source-name*, where table-name is the name of the table in the global schema and source-name is the name of the data source. It supports SQL-like queries that combine data in relational data stores and data in Bigtable data stores for cloud system. Bigtable is accessed through the Google Query Language (GQL). The system relies on mapping a limited set of relational operations to native queries expressed in GQL. However, it only works for Bigtable-like systems and cannot integrate data from other families of NoSQL systems, such as document or graph databases. The data model used is relational which is specific solution that reduces the scope of application.

QoX [26] is a Multi-store system for data warehouse systems that integrates data from relational databases, and others execution engines such as MapReduce or Extract-Transform-Load (ETL) tools. The data flow combine unstructured data with structured data and use both generic data flow operations like filtering, join, aggregation. The QoX Optimizer uses xLM, a proprietary XML-based language to represent data flows. It uses some appropriate wrappers to translate xLM to a tool-specific XML format and vice versa, the QoX Optimizer may connect to external ETL engines and import or export dataflows. This work focuses on the proposal of algorithms for data flow optimization. However, the use of a proprietary language (xLM) allows only the translation to XML, other languages cannot be managed. This is also due to the levels of abstraction dealt with, namely logical and physical levels, which does not allow to manage other kinds of data stores.

Polybase [9] is a multi-store system of the SQL Server Parallel Data Warehouse (PDW). It allows users querying unstructured data stored in a Hadoop cluster (HDFS) using SQL and integrate them with relational data in the store. The HDFS data is referenced in Polybase as external tables making the correspondence with the HDFS file on the Hadoop cluster, and manipulated using SQL queries. The query optimizer uses SQL operators on HDFS data to translate queries into MapReduce jobs to be executed directly on the Hadoop cluster. The major drawback is that it allows the integration of RDBMS with HDFS only.

FORWARD [21] is a multi-store system that supports SQL-like language (SQL++) designed to unify the data model and query language capabilities of NoSQL and relational databases. Forward multi-store system uses the GaV

(Global as View) approach, where each data source appears as an SQL++ virtual view, defined over SQL++ collections. SQL++ extends both the JSON and relational data models. Here, the classical SELECT statement of SQL is adapted and extended to perform queries on collections of JSON objects. However, it does not deal explicitly with heterogeneity of objects, since it does not provide constructs similar to the WHERE branches. Furthermore, there is a need of temporary persistent database in order to save intermediates results during the execution of complex transformations requiring several queries executed sequentially.

BigDAWG [11] is a multi-store system, with the goal of unifying querying over a variety of data models and languages. A key user abstraction in BigDAWG is an island of information, which is a collection of data stores accessed with a single query language. It can be a variety including relational (RDBMS), Array DBMS, NoSQL and Data Stream Management System (DSMS). The island allows a wrapper to map the island language to its native one. For queries that accesses more than one data store, the CAST operation is used to copy objects on data stores. Each query, that involve multiple islands, must be expressed as multiple sub-queries, each one in a specific island language. The user encloses each sub-query in a SCOPE specification. Thus, a multi-island query will have multiple scopes to indicate the expected behavior of its sub-queries. Note that SCOPE and CAST operations are specified by the user. Although, BigDAWG polystore contextualizes queries and cast data between multiple stores. However, there is a need for human intervention to manually introduce the scope clauses, specifying the system for executing portions of an algorithm, and cast clauses for moving data between system, thus employing the best data store for a particular task.

All aforementioned works aspire to propose a system that manages different data management systems (such as RDBMS, NoSQL, XML, HDFS, ...), in order to deal with variety data sources and resolve the problem of accessing heterogeneous data. Finding, existing polystore systems provide integrated access to a number of data stores through one or more query languages with a logical common data model, generally relational. However, several multi-store systems are being built for specific systems, mainly due to the abstraction level of the design of global schema (logical or physical). Besides, they are defined with different objectives (unifying schema, querying, performance, etc.), query processing approaches, architectures, etc. which make hard to choose the appropriate one for the deployment of systems. There is still no generic solution to manage different data stores systems unifying languages and variety of data sources.

A comparison of listed multi-store systems is given in Table 2 based on five criteria: (1) The type of data stores involved (RDBMS, NoSQL, MapReduce, HDFS, ETL Tools, Array DBMS ...); (2) the goal of work (Querying, Unifying of schema or Performance of ETL data flows), (3) Query languages used (SQL, SQL++, SQL-like, ...); (4) Perspective of design: Conceptual (C), Logical (L), Physical (P); and (5) Usage of common data models for each data store.

Table 2. Comparison of multi-store systems

Work	Criteria				
	Data stores	Goal	Language	Design	Model
BigIntegrator [39]	RDBMS/BigTable	Querying	SQL-Like	P	Relational
QoX [26]	RDBMS /MapReduce ETL Tools	Performance	XML	L/P	Graph
Polybase [9]	RDBMS/HDFS	Querying	SQL	P	Relational
FORWARD [21]	RDBMS/NoSQL	Unifying	SQL++	P	JSON
BigDAWG [11]	RDBMS/NoSQL/ Array/DSMSs	Unifying	CAST/SCOPE	P	/

3   ETL Workflow Meta Model

The purpose of this section is to present a meta model of ETL workflow that allows a generic modeling of the activities of the ETL environment.

An ETL workflow is a collection of activities and transitions between activities. An activity correspond to one ETL scenario and a transition determine the execution sequence of activities to generate a data flow from sources to target stores. An activity involves elements of data sources, elements of data stores and ETL operators allowing the definition of expressions, characterizing the semantics of the data pushed to the target schema. We assume that ETL workflow consists of n sequential activities $(A_1 \ldots A_n)$. Each activity is either elementary, can not be decomposed, or complex. A complex activity consists of a set of parallel and sequential elementary activities. Figure 3 shows examples of ETL operator, activity and workflow definition. The illustrated example of the ETL activity considers students taking a given course.

Based on the WfMC[1] definition of the workflow Meta model, we propose a Meta model for the ETL workflow that deals with the complexity of data sources and stores heterogeneity and captures the data semantics in the context of Data warehouses.

The Meta model, depicted in Fig. 4, involves the particular scenarios applicable to a generic data warehouse. *ETL flow* entity in the Meta model is the main entity having a reflexive relation that captures the sequence of events producing the ETL process. ETL flow is composed by a set of activities. An *activity* represents a unit of work which involve a set of elements from input data sources schemata, a set of elements from data stores output schemata and a set of ETL operators, so that the activity is populated each time with its proper parameter values. Elements may be entities, properties or even constraints, captured by the entity *Element*. An example of a simple ETL activity can be to sort student entity based on Age property. An activity can be complex, as captured by the reflexive composite relation of Activity entity, defining composite activities. With a physical perspective, an activity corresponds to an execution of

---

[1] http://www.wfmc.org/.

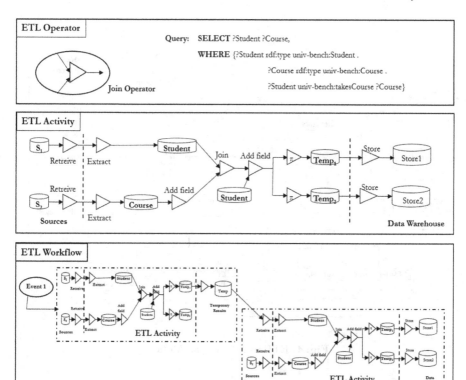

Fig. 3. ETL operator, ETL activity and ETL workflow.

one job. The ETL flow of data from sources towards target stores is done across a sequencing of the activities achieved through a transition from one activity to another, defined by the *Transition* entity. A transition is employed for the interconnection of activities participating in the ETL process, for generating a data flow.

3.1   Overloading: Example of Sorting an Array with C++

In this section, we describe the notions of overloading and genericity features of the programming languages by giving an example on sorting an array having elements with different data types. Here, we aim to point out the interest of mapping these notions to our problematic, namely the data sources integration and polystore deployment. On the basis of this principle, the *types of data*, such as: integer, string, etc., and the *functions* of programming languages respectively correspond to the kinds of *data sources* to be integrated and the *ETL operators*.

In programing languages, the overloading occurs when two different functions have the same name and their parameters are different; either because they have a different number of parameters, or because any of their parameters are of a

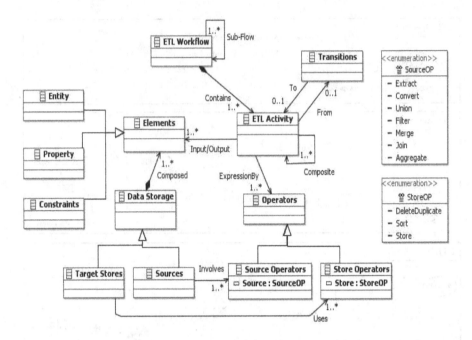

Fig. 4. ETL workflow MetaModel.

different type. The compiler knows which one to call in each case by examining the types passed as arguments when the function is called. For example, if it is called with two int arguments, it calls to the function that has two int parameters, and if it is called with two string, it calls the one with two string. For cases such as this, C++ has the ability to define functions with generic types, known is *function templates*. The template parameters, having the syntax: template<class ItemType>, can be generic template types by specifying either the class or typename.

In the case where the name of an overloaded function template is called, the compiler will try to deduce its template arguments and check its explicitly declared template arguments. If successful, it will instantiate a function template specialization, then add this specialization to the set of candidate functions used in overload resolution. The compiler proceeds with overload resolution, choosing the most appropriate function from the set of candidate functions.

Now, we describe an example of sorting an array having elements with different types of data. In C++, instead of writing a different sorting subroutine for each type of data, we can write a single subroutine template. The template is not a subroutine; it is a factory for making subroutines. To better appreciate how template functions are developed, consider the following implementation of selection sort using C++. Note that the implementation implicitly assumes the existence of a copy constructor, an assignment operator, and a comparison operator (<) for ItemType.

```
template <typename ItemType>

void Swap(ItemType& a, ItemType& b) {
  ItemType tmp = a;        // copy constructor
  a = b;                   // assignment operator
  b = tmp;          // assignment operator
}

template <typename ItemType>

unsigned FindMin(const ItemType A[ ], unsigned begin, unsigned end) {
  assert(begin <= end);
  unsigned m = begin;
  for (unsigned i = begin + 1; i < end; i++)
    if (A[i] < A[m])       // comparison operator
      m = i;
  return m;
}

template <typename ItemType>

void Sort(ItemType A[ ], unsigned N) {
  for (unsigned i = 0; i < N; i++) {
    unsigned m = FindMin(A, i, N);
    Swap(A[i], A[m]);
  }
}
```

To instantiate the Sort template so that it sorts *int* arrays in descending order, we can implement the following:

```
bool greater_than(const int& a, const int& b) { return a > b; }
...
Sort<int, greater_than >(...);
```

More interestingly, one could also reuse the Sort template in ascending order, to sort arrays of C *strings*:

```
typedef const char *CString;
bool c_string_less_than(const CString& x, const CString& y) {
  return strcmp(x, y) < 0;
}
...
Sort<CString, c_string_less_than>(...);
```

Note that the C++ compiler generates a version of the code for each type used in the program, in a process known as *instantiation* of template. For example, with type of int, the following code will be generated by the compiler for the swap function:

```
void swap(int &a, int &b) {
int temp;
temp = a;
a = b;
b = temp;
}
```

4   Background and a Motivating Example

In this section, we give an overview on the most important types of databases: relational, semantic and graph databases adopted by many sources. Then, a motivating example is considered to illustrate the basic ideas behind our proposal.

4.1   RDF and SPARQL

RDF[2] is a set of 4-tuple $<s, o, p, g>$ forming an RDF triple where its subject s has the property p, and the value of that property is the object o, and the graph label IRI g. Note that the graph label g can be omitted, in which case the triples are considered part of the default graph of the RDF data-set. N-Quads is a line-based, plain text format for encoding an RDF dataset. It defines an RDF dataset composed of RDF graphs composed of a set of RDF triples.

Definition 1. *RDF Graph: denoted as g, is a finite set of RDF N-Quads in which every 4-tuple describes a directed edge labeled with p from the node labeled with s to the node labeled with o belonging to the graph g. Subjects s can be URIs or blank nodes, properties p are URIs, while objects o can be URIs, blank nodes, or literals (i.e., values).*

SPARQL[3] is the standard query language for RDF. Similarly, SPARQL query can also be represented as a query graph.

Definition 2. *SPARQL Query Graph: denoted as q, is a finite set of RDF triple patterns; some of the nodes in a pattern are variables which may appear in multiple patterns.*

An example of RDF and SPARQL graphs using LUBM Benchmark are respectively depicted in Fig. 5(a) and (b).

---

[2] https://www.w3.org/RDF/.
[3] www.w3.org/TR/rdf-sparql-query/, 2008.

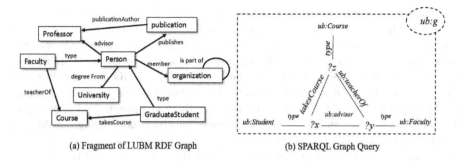

(a) Fragment of LUBM RDF Graph   (b) SPARQL Graph Query

Fig. 5. Example of RDF and SPARQL graphs.

4.2 Formalization of Databases

A relational database is defined by set of tables, attributes, instances and constraints. Consequently, each source S_i is defined by set of entities $E = \{e_1, e_2, \ldots, e_m\}$, a set of attributes related to entities $A = \{a_1, a_2, \ldots, a_m\}$ and a set of record-sets $R = \{r_1, r_2, \ldots, r_n\}$. The ETL process is formalized as follows:

$$\mathcal{ETL}(\mathcal{E}, \mathcal{A}, \mathcal{R}) : <\mathcal{G}_{(\mathcal{E},\mathcal{A})}, \mathcal{S}_{(\mathcal{E},\mathcal{A},\mathcal{R})}, \mathcal{M}^{(\mathcal{E},\mathcal{A})}_{(\mathcal{G},\mathcal{S})}>,$$

where the $\mathcal{G}_{(\mathcal{E},\mathcal{A})}$ is the global schema defined using the alphabet of the domain entities and attributes (relation for relational, class for object-oriented, etc.), $\mathcal{S}_{(\mathcal{E},\mathcal{A},\mathcal{R})}$ is the data sources that the schema is derived from the alphabet of the domain entities and attributes and populated using record-sets R, $\mathcal{M}^{(\mathcal{E},\mathcal{A})}_{(\mathcal{G},\mathcal{S})}$ is the mappings made by a set of assertions using a query language over the entities and attributes alphabets of $\mathcal{G}_{(\mathcal{E},\mathcal{A})}$ and $\mathcal{S}_{(\mathcal{E},\mathcal{A},\mathcal{R})}$. An example of the instantiation of the formalization is given:

$\mathcal{ETL}(\mathcal{E}, \mathcal{A}, \mathcal{R}) : <$Relational schema ($Student_G(name)$, $university_G(name)$, $courses_G(title)$), Oracle relational databases ($Student_S$ (name:#Student1, name:#student2), $university_S$ (name:#univ1, name:#univ2)), mappings between table attributes ($Student_G.name = Student_S.name>$.

A semantic database (*SDB*) is formally defined as follows [5]:
   $<OM, I, Pop, SL_{OM}, SL_I>$, where:

- *OM*: $<C, R, Ref, formalism>$ is the ontology model of the *SDB*; where C and R denote respectively concepts and roles of the model; Ref is a function defining terminological axioms of a DL TBOX (Terminological Box) [3], (e.g., Ref(Student) $\rightarrow$(Person $\sqcap \forall$ takesCourse(Person, Course))) and *Formalism* is the formalism followed by the global ontology model like RDF, OWL, etc.);
- *I*: presents the instances (the ABox) of the *SDB*;
- *Pop*: C $\rightarrow 2^I$ is a function that relates each concept to its instances;
- SL_{OM}: is the Storage Layout of the ontology model (vertical, binary or horizontal) [14]; and
- SL_I: is the Storage Layout of the instances I.

The ETL process is defined as follows: $\mathcal{ETL}(\mathcal{IO}, \mathcal{SDB})$ $<\mathcal{G}, \mathcal{SDB}, \mathcal{M}>$,

where the $\mathcal{G}$ is the global schema defined by an integrated ontology (IO) containing all concepts and properties satisfying user requirements. Three scenarios are possible: G = IO: the IO corresponds exactly to users requirements, G $\subset$ IO: the DISO is extracted from the IO, G $\supset$ IO: the IO does not fulfil all the requirements. The designer extracts the fragment of the IO corresponding to the requirements of the $\mathcal{G}$ and enriches it with new concepts and properties. The mappings assertions between $\mathcal{SDB}$ sources and IO is done using ontological constructors defined in our previous work [5].

In our case, we consider the first scenario (G = IO). The ETL process is applied in order to populate IO with available instances extracted from sources.

An example is done as follows: $< \mathcal{G}$: LUBM Ontology, $\mathcal{SDB}$s: Oracle $\mathcal{SDB}$s, $\mathcal{M}$: mappings assertion between local ontologies of sources and LUBM ontology$>$.

A graph database usually used to represent knowledge bases through a graph G whose nodes (V), edges (E) and labels (L_v, L_e) represent respectively classes, instances and data properties, object properties and **DL** constructors. Neo4J[4] is an example of a storage system of graph databases.

Definition 3. *Property Graph is defined in [23] as a directed, labelled, attributed, multi-graph. Graphs of this form allow for the representation of labelled vertices, labelled edges and attribute meta-data (properties) for both vertices and edges. The other types of graphs can be obtained from graph property structure by adding or abandoning particular characteristics.*

For example, restricting the vertex/edge labels to Uniform Resource Identifiers (URIs) generates a Resource Description Framework (See footnote 2) (RDF). The ETL process is formalized as a Directed Acyclic Graph (DAG) noted G, consisting of a set of vertices (V), a set of edges (E) and a set of labels (L), where:

ETL : $<(V, E, L)_I, (V, E, L)_O, Op, Function, ETLGraph>$, where:

- $(V, E, L)_I = \{(v, e, l)_1, (v, e, l)_2, ,(v, e, l)_n\}$ finite set of input labeled vertices/edges, where $E \subseteq N \times N$;
- $(V, E, L)_O = \{(v, e, l)_1, (v, e, l)_2, ,(v, e, l)_n\}$ finite set of output labeled vertices/edges, where $E \subseteq N \times N$;
- *Op*: is a set of operators commonly encountered during ETL process (merge, union, join, etc.);
- *Function*: is a function over a subset of Input-Set nodes applied in order to generate data satisfying restrictions defined by ETL operators;
- *ETLGraph*: is a set of output nodes (concepts and instances) added to final $\mathcal{DIS}$ schema and linked by edges.

---

[4] https://neo4j.com/product/.

4.3   Motivating Example

To explicit the basic ideas behind our proposal, let us consider a scenario, where a governmental organisation wants constructing a data warehouse to analyse the performance of students in universities. To do so, this organisation considers four data sources with a high variety. The particularity of these sources is that they are derived from the benchmark related to the universities (LUMB[5]) and the Yago[6] knowledge base. The details of these sources are given below: S_1 is a MySQL relational databases with the following schema composed of tables and attributes: *Student(name), Course(title), University*, S_2 is a Berkeley XML *DB* with a schema composed of elements and attributes: *GraduateStudent(name), GraduateCourse(title), University*, S_3 is an Oracle RDF *SDB* composed of classes, properties: *Student(name), Publication, University*, and S_4 is a Neo4j Graph *DB* with nodes, edges: *Person, Student(name), Publication, PublicationAuthor, University*.

The obtained warehouse has two stores Semantic Oracle and Mongodb. In this context, the different ETL operators have to be overloaded to deal with this variety. Figure 6 describes the whole architecture of the ETL process, where *Extract* and *Convert* operators are overloaded. As we see, they have the same name, but different signatures. Based on the format of each store, the Store operator is also overloaded.

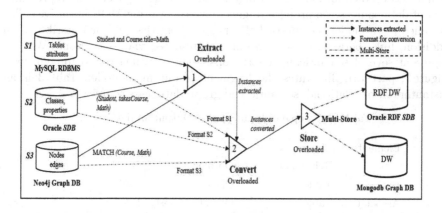

Fig. 6. An example of ETL operator overloading

5   Generalisation of ETL Elements

Before discussing our proposal in overloading ETL operators, we first formalize the ETL process and its operators. An ETL process is defined as 5-tuples as follows: $<InputSet, OutputSet, Operator, Function, ETLResul>$, where:

---

[5] http://swat.cse.lehigh.edu/projects/lubm/.

[6] www.yago-knowledge.org/.

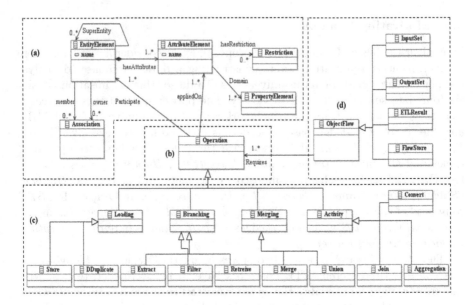

Fig. 7. Excerpt of ETL meta models

InputSet: represents a finite set of input elements describing data sources. Each source has its own format and storage layout. To make generic the representation of data sources, we propose to generalize them using MOF initiatives. The obtained meta-model is composed of conceptual entities and their attributes. In addition, links between entities are also represented via associations. We also represent several semantically restrictions, such as primary and foreign keys. Figure 7, part (a), illustrates the fragment of our meta-model. Table 3 is an instantiation of relational, semantic and graph databases sources.

Table 3. Sample of InputSet and OutputSet databases.

Elements	Databases		
	Relational	Ontological	Graph
Entity	Table	Class	Node
Association	Table	Object_property	Edge
Attribute	Column	Data_property	Node
Property	Domain of values	Data type	Domain of values
Restriction	Primary key	SameAs	Node

OutputSet: is a finite set of intermediate or target elements. The output of the ETL process can be either the intermediate output (sub process) or the final output (ETL process). The final output corresponds to the target data stores, where the schema of each store can be seen as an instance of our meta-model (part (a) of Fig. 7).

Operator: is a set of operators commonly encountered during the ETL process in [28]. By analysing these operators, we propose to decompose them into four categories: (1) loading class, (2) branching class, (3) merging class and (4) activity class.

- **Branching Class**: delivers multiple output-sets which can be further classified in Filter operations based on conditions or Extract and Retrieve operations that handle with the appropriate portion of selected data.
- **Merging Class:** fuses multiple data incoming from data sources. We identify two possible operations: (i) Merge operation applied when the data belong to attributes related to entities of the same source; (ii) Union operation applied when data belong to entities incoming from different data sources
- **Activity Class:** represents points in the process where work is performed. It corresponds to all operations of join conversion and aggregation. Join operations is applied when data belong to different entities. Conversion operation is applied on data having different format in order to unify it and adapt it to the target data stores. The aggregation operation is done depending on the schema of the target data stores applying needed functions (count, sum, avg, max, min).
- **Loading class:** represents the point of data quality by the detection of duplicated data and cleaning them before their loading in the target data store.

Function: is a function over a subset of Input-Set applied in order to generate data satisfying restrictions defined by any ETL operator.

ETLResult: is a set of output elements representing the flow.

Based on this, we propose a meta model of these operations (part (c) of Fig. 7).

The generic formalization of each operator is given by:

- $Retrieve(S, E, A, R)$: retrieves data D of attributes A related to entities E from Source S;
- $Extract(S, E, A, R, CS)$: enables the data D extraction of A related to entities E from source S satisfying constraint CS;
- $Merge(S, E_1, E_2, A_1, A_2, D_1, D_2)$: merges data D_1 and D_2 belonging to the source S;
- $Union(S_1, S_2, A_1, A_2, D_1, D_2)$: unifies data D_1 and D_2 belonging to different sources S_1 and S_2 respectively;
- $Filter(S, E, A, D, CS)$: filters incoming data D, allowing only values satisfying constraints CS;
- $Join(S, E_1, E_2, A_1, A_2, D_1, D_2)$: joins data D_1 and D_2 having common attributes A_1 and A_2;
- $Convert(S, E, A, D, F_S, F_T)$: converts incoming data D from the format F_S of source S to the format of the target data store F_T;
- $Aggregate(S, E, A, D, F)$: aggregates incoming data D applying the aggregation function F (count, sum, avg, max) defined in the target data-store.
- $DD(D)$: detects and deletes duplicate values on the incoming data D;
- $Store(T, E, A, D)$: loads data D of attributes A related to entities E in the target data store T,

---

5.1   Instantiation of the ETL Meta Model

The concepts of the ETL process meta model are categorized into different aspects namely: (a) Basic concepts derived from MOF M3 layer, it enable the modeling of all the data sources participating in the integration process; (b) functional aspect: supports the definition of operations (functions and proce-dures with membership constraints based on parameters); (c) behavioural aspect: required to demonstrate the ETL flow using the generic ETL operators and (d) informational aspects that represents the inputs and outputs flow of the ETL process as well as the results and the target data stores. The different parts (b) (c) and (d) remain unchanged regardless of the type of data sources and stores. The part (a) is instantiated for each type of data source: Relational, ontological and graph databases (Fig. 8).

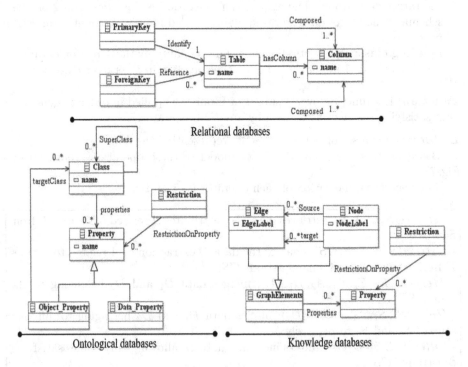

Fig. 8. Instantiation of the meta-model (part a) for Relational, Ontological and Graph data sources.

5.2   Overloading Operators

In this section, we show the mechanism to overload ETL operators by considering semantic and graph databases.

In the case of a semantic database, the signature of overload operators is as follows:

- $Retrieve(S_i, C, I)$: retrieves instances I related to classes C from Source S_i;
- $Extract(S_i, C, I, CS)$: extracts instances I related to classes C from source S satisfying constraint CS;
- $Merge(S_i, C_1, C_2, I_1, I_2)$: merges instances I_1 and I_2 related to the classes C_1 and C_2 respectively and belonging to the same source S_i;
- $Union(S_1, S_2, C_1, C_2, I_1, I_2)$: unifies instances I_1 and I_2 related to C_1 and C_2 respectively and belonging to different sources S_1 and S_2 respectively;
- $Filter(S_i, C, I, CS)$: filters incoming instances I related to C, allowing only the values satisfying constraints CS;
- $Join(S_i, C_1, C_2, I_1, I_2)$: joins instances I_1 and I_2 related to C_1 and C_2 respectively and having common object properties;
- $Convert(S_i, C, I, F_S, F_{TDW})$: converts incoming instances I from the format F_S of source S_i to the format of the target TDW (F_{TDW});
- $Aggregate(S, C, I, F)$: aggregates incoming instances I related to C applying the aggregation function F (count, sum, avg, max) defined in the TDW.
- $DD(I)$: detects and deletes duplicate values on the incoming instances I;
- $Store(TDW, I)$: loads instances I in target TDW.

In the case of a graph database, the signature of overload operators is as follows:

- $Retrieve(G, V_j, L_j)$: retrieves a node V_j having an edge labeled by L_j of G.
- $Extract(G, V_j, CS)$: extracts, from G, the node V_j satisfying CS.
- $Convert(G, G_T, V_i, V_T)$: converts the format of the node V_i to the format of the target node V_T. The conversion operation is applied at instance level.
- $Filter(G, V_i, CS)$: applied on V_i node, allowing only instances satisfying CS.
- $Merge(G, V_i, V_j)$: merge instances denoted by nodes V_i, V_j in same graph G.
- $Union(G, G_T, V_i, V_j, E_j)$: links nodes that belongs to different sources. It adds in the target graph G_T, both nodes V_i and V_j and link them by an edge E_j.
- $Join(G, G_T, V_i, V_j, E_j)$: joins instances whose corresponding nodes are $V_i \in G$ and $V_j \in G_T$. They are linked by an *object property* defined by an edge E_j.
- $Store(G_T, V_j)$: loads instances denoted by nodes V_j to the target graph G_T.
- $DD(G_T, CS)$: sorts the graph G_T based on CS and detects duplication.
- $Aggregate(G_T, V_j, Op)$: aggregates instances represented by the nodes V_j.

Some primitives need to be added to manage the ETL operations required to build the ETLgraph such as:

- $AddNode(G_T, V_j, E_j, L_j)$: adds node V_j, edge E_j, label L_j required to G_T.
- $UpdateNode(G_T, V_j, E_j, L_j)$: updates node V_j, edge E_j, label L_j in G_T.
- $RenameNode(G_T, V_j, E_j, L_j)$: renames node V_j, edge E_j, label L_j in G_T.
- $DeleteNode(G_T, V_j, E_j, L_j)$: deletes node V_j, edge E_j, label L_j from G_T.
- $SortGraph(G_T, V_j, CS)$: sorts nodes of G_T based on some criteria CS to improve search performance.

Our goal is to facilitate, manage and optimize the design of the ETL process during the initial design, deployment phase and during the continuous evolution of the $\mathcal{TDW}$. For that, we enrich the existing ETL operators with *split*, *context* and *Link* operators elevating the clean-up and deployment of ETL process at the conceptual level.

- $Split(G, G_i, G_j, CS)$: splits G into two sub-graphs G_i and G_j based on CS.
- $Link(G_T, V_i, V_j, CS)$: links two nodes V_i and V_j using the rule CS.
- $Context(G, G_T, CS)$: extracts from the graph G a sub-graph G_T that satisfies the context defined by restrictions CS using axioms.

6   Deployment on a Multistores System

In this section, we propose a methodology to satisfy the $n-m$ scenario discussed in the Introduction. To do so, we have to consider three issues: consolidation of schemas, fusion of instances, and deployment.

6.1   Consolidation ETL Algorithm

Algorithm 1 describes in details the overloading of ETL operators in the context of semantic and graph data sources. It is based on mappings defined between data sources schemes and global schema. We used mappings described in [5].

6.2   Fusion Procedure

In this section, we propose a fusion method to merge different input data sources representations based on the target model chosen by the designer. Our solution is based on the *Graph Property* model presented above [23]. The property graph is common because modellers can express other types of models or graphs by adding or abandoning particular elements. To do so, we propose to use the primitives proposed previously. They enable designers to adds, deletes

Algorithm 1. Overloading ETL Process Algorithm

Input: $\mathcal{IO}$ or Contextual $\mathcal{KB}$, S_i: Local sources $\mathcal{SDB}$
Output: $\mathcal{TDW}$ (schema + instances)

1: $V_{Si} := \emptyset$; $\text{ETL}_G := \text{Graph}(\text{Tbox}(kb))$; $V_{kb} := \text{GetNodes}(kb)$;
2: **if** Input is $\mathcal{IO}$ **then**
3:    $Input_{cond} := (\text{C} : \text{Class of ontology IO})$;
4: **else if** Input is $\mathcal{KB}$ **then**
5:    $Input_{cond} := (V_i \in V_{kb} \wedge (V_i \text{ isClass}))$;
6: **end if**
7: **for** each $Input_{cond}$ **do**
8:    **for** Each S_i **do**
9:       **if** *Equivalent or complete mappings* $(N_{S_i}, N_{KB}) \vee (C_{S_i}, C_{IO})$ **then**
10:          **if** Input is $\mathcal{IO}$ **then**
11:             $\text{C} := \text{IdentifyClass}(C_{TDW}, \text{C}_i)$;
12:          **else if** Input is $\mathcal{KB}$ **then**
13:             $V_i := \text{IdentifyNode}(\text{ETL}_G, V_i)$;
14:             $E_i := \text{IdentifyEdge}(\text{ETL}_G, E_i)$;
15:          **end if**
16:       **else if** *sound or overlap mappings* $(N_{S_i}, N_{KB}) \vee (C_{S_i}, C_{IO})$ **then**
17:          **if** Input is $\mathcal{IO}$ **then**
18:             $\text{Const} := \text{ExtractConstraint}(C_{TDW}, \text{C}_i)$;
19:          **else if** Input is $\mathcal{KB}$ **then**
20:             $\text{Const} := \text{ExtractNeighbor}(\text{ETL}_{Graph}, V_i)$;
21:          **end if**
22:       **end if**
23:       **if** (Input is $\mathcal{IO}$) **then**
24:          **if** (Const isDataTypeProperty) **then**
25:             $\text{I} := \text{Convert}(\text{C}_j, \text{I}, \text{const})$;
26:             $\text{I} := \text{Filter}(\text{C}_j, \text{I}, \text{Const})$;
27:          **else if** (Const isObjectProperty) **then**
28:             $\text{I} := \text{Join}(\text{C}_j, \text{C}_i, \text{I}, \text{Const})$;
29:          **else if** (Const isAxiom) **then**
30:             $\text{I} := \text{Aggregate}(\text{C}_j, \text{C}_i, \text{I}, \text{Const})$;
31:          **end if**
32:          $\text{I} := \text{MERGE}(\text{C}_{S_i}, \text{I})$; $\text{I} := \text{UNION} (\text{C}_{S_i}, \text{C}_{IO}, \text{I})$;
33:          $\text{STORE}(\text{IO}, \text{C}_i, \text{DD}(\text{I}))$;
34:       **else if** (Input is $\mathcal{KB}$) **then**
35:          **if** (Const isDataTypeProperty) **then**
36:             $\text{I} := \text{Convert}(V_j, \text{I}, \text{const})$;
37:             $\text{I} := \text{Filter}(V_j, \text{I}, \text{Const})$;
38:          **else if** (Const isObjectProperty) **then**
39:             $\text{I} := \text{Join}(V_j, V_i, \text{I}, \text{Const})$;
40:          **else if** (Const isAxiom) **then**
41:             $\text{I} := \text{Aggregate}(V_j, V_i, \text{I}, \text{Const})$;
42:          **end if**
43:          $\text{I} := \text{MERGE}(N_{S_i}, \text{I})$; $\text{I} := \text{UNION} (N_{S_i}, N_{kb}, \text{I})$;
44:          **for** Each I **do**
45:             $\text{ETL}_G := \text{addEdge}(\text{ETL}_G, N_i, \text{edge}, \text{I})$;
46:             $\text{ETL}_G := \text{addNode}(\text{ETL}_G, N_i, \text{edge}, \text{I})$;
47:          **end for**
48:          $\text{ETL}_G := \text{Filter}(\text{ETL}_G, \text{DD}(N_i), \text{Null-values})$;
49:          $\text{STORE}(\text{ETL}_G)$;
50:       **end if**
51:    **end for**
52: **end for**

and renames graph elements in order to manage the ETL flow generated and adapt it to the target storage layout chosen. An example of *addnode* primitive is done as follows:

```
- Sparql query language :
construct ?V
where {GRAPH  :?G {?V rdf:type name-space:Class}}
- Using Cypher query language for Neo4j graph database:
MERGE (<node-name>:<label-name>
{<Property1-name>:<Pro<rty1-Value> ..... <Propertyn-name>:<Propertyn-Value>}
```

On the basis of items presented previously, we have identified three particular cases:

Deployment of $\mathcal{KB}$ on $\mathcal{SDB}$: the RDF graph allowing the representation of $\mathcal{KB}$ deployed on $\mathcal{SDB}$ can be obtained by restricting labels of the nodes and edges to Uniform Resource Identifiers (URIs) and not allowing node/edge attributes;

Deployment of $\mathcal{KB}$ on graph database: using graph property having directed, labelled, attributed nodes and edges will allow a deployment of $\mathcal{KB}$ on a graph system;

Deployment of traditional data on graph: starting from a property graph, we generate a standard semantic graph by discarding the nodes/edges attributes. Having a semantic graph, we consider the nodes as attributes/data of traditional data, labels nodes are either *attributes or data*, edges as relationships between data and attributes of traditional data, labels edges can be either *has_data or has_attributes*.

6.3   Deployment of ETL Process

Storage deployment models can follow different representations according to specific requirements. A $\mathcal{TDW}$ can be deployed using horizontal, vertical, hybrid models, NoSQL, etc. [14]. In our case, we choose to deploy the $\mathcal{TDW}$ into vertical representation using Oracle DBMS which offers a storage model to represent instances and graphs using Oracle RDF Semantic Graph. We translated the $\mathcal{TDW}$ schema into vertical relational model, then generated an N-Triple file, load it into a staging table using Oracle's SQL*Loader utility. We applied the ETL Algorithm to populate the target schema.

The proposed tool is implemented in Java language and uses JENA API to access ontologies and a $\mathcal{KB}$. Each generic ETL operator is implemented as a Web Service Restful using Java overload polymorphism implementation. The restful web service is implemented is such way to consider the overload resolution. Each ETL operator is overloaded by determining the most appropriate definition to use. It compares the argument type used to call the appropriate service restfull with the parameter types specified in the definitions. This will allow managing the different representations of input data (instances and graph) by overloading each ETL operator according to input representation.

The proposed ETL algorithm consists then in orchestrating the Web services. Based on the existing mappings between the schemes of sources and the target DW schema, the implemented Web services allows an automatic extraction of the appropriate data from the sources, their transformation (filtering, conversion and aggregation) and the computation of the new values obeying to the structure of the input data and target one. The storage deployment of the DW is done according to the target platform. The diversity of storage models (vertical, horizontal, and hybrid) is handled by our proposed tool, where the suitable web service is invoked in order to translate the logical schema according to the physical model of the target DBMS. Each Web service that accesses the persistent storage is implemented using Data Access Object (DAO) Design patterns[7]. DAO implements the access mechanism required to handle the different input representations. The DAO solution abstracts and encapsulates all access to persistent storage, and hides all implementation details from business components and interface clients. The DAO pattern provides flexible and transparent accesses to different storage layout. In order to obtain a generic implementation of the ETL process, we implemented our solution following service oriented architecture (SOA). SOA offers the loose coupling of the web services defined bellow, and interaction among them. It allows the integration of new web services without affecting the existing one. This provides the flexibility of the input data and physical deployments of DW.

The application implements an orchestration of web services in early binding. Indeed, each web service is implemented in such way that parameters and variables are detected and checked at compile time. Figure 9 describes the whole architecture of the ETL and MultiStore Services. A demonstration video summarizing the different services offered by our proposal is available at: https:// youtu.be/zbtl1qMvPOU.

7   Experimental Study

In order to illustrate the feasibility of our approach, we use our motivating example (cf. Sect. 3). We choose Oracle semantic database system to implement the sources and the warehouse. Oracle 12c release 2 delivers *RDF Semantic Graph features* as part of Oracle Spatial and Graph. With native support for RDF and OWL standards for representing semantic data, with SPARQL for query language. Oracle has defined two subclasses of DLs: OWLSIF and a richer fragment OWLPrime. We use the OWLPrime fragment which offers the following constructors: *rdfs:domain,rdfs:range, rdfs:subClassOf, owl:sameAs, rdfs:subPropertyOf, owl:equivalentClass, owl:equivalentProperty, owl:inverseOf, owl: TransitiveProperty, owl:SymmetricProperty, owl: FunctionalProperty, owl: InverseFunctionalProperty.* Note that *OWLPrime* limits the expressive power of DL formalism in order to ensure decidable query answering. We have deployed the $SDBs$ created into Oracle semantic database system. Note that the source S1 has been created from the LUBM Ontological schema and populated using UBA

---

[7] http://www.oracle.com/technetwork/java/dataaccessobject-138824.html.

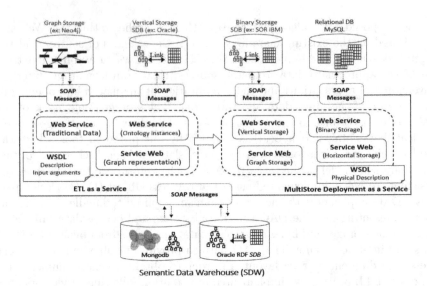

Fig. 9. A general architecture of the ETL and MultiStore Services.

generator (class and properties structure). While sources S2 and S3 have been created and populated using YAGO knowledge base (RDF graph structure). The proposed Algorithm 1 was implemented using the overload of ETL operators in order to integrates the created sources into the DW taking in account their heterogeneity. Note that generic ETL operators defined in the previous section are expressed on the conceptual level. Therefore, each operation has to be translated according the logical level of the target DBMS (Oracle). Oracle offers two ways for querying semantic data: SQL and SPARQL. We choose SPARQL to express this translation. Here an example of $\mathcal{KB}$ aggregation ETL operator translation to SPARQL:

```
PREFIX yago: http://yago-knowledge.org/resource/yago.owl#
AGGREGATE: Aggregates incoming record-set.
 Select (Count(?Instance) AS ?count) Where {
GRAPH   :?G {?Instance rdf:type yago:Class}}
Group By ?Instance.

Example: universities per country
Select (count(?university) AS ?count, ?city) Where {
GRAPH   :ETLGraph {?university rdf:type yago:University . yago:city ?name .
FILTER (?name = 'France')} }
Group By ?university.
```

The translation of CONTEXT operator to SPARQL query generates the ETL Graph from YAGO $\mathcal{KB}$ using yagoWordIn Context object property as predicate, university value as object, and subject being a class of resources. It will provide an additional information required for RDF data analysis.

```
PREFIX  rdf:<http://www.w3.org/1999/02/22-rdf-syntax-ns#>
PREFIX  yago: http://yago-knowledge.org/resource/yago.owl#
CONSTRUCT
{SELECT   ?g ?x ?y
WHERE  {
GRAPH:ETLGraph { ?x ?y ?z . Filter
(?x rdf:type owl:DatatypeProperty .
x? yago:hasContext yago:university .
OPTIONAL  x? yago:yagoWordInContext yago:university)
}}
GROUP  BY ?g ?c ;}
```

7.1   Evaluation Study

In this section, we present the performance of our approach through a set of experiments considering an Ontology and large $\mathcal{KB}$. Four criteria are used to evaluate our proposal: (i) complexity of the proposed ETL algorithm, (ii) evaluation time per ETL operators before and after overloading, (iii) scalability of the ETL process, (iv) inference performance.

Environment of our Experiments. Our experiments are based on LUBM ontology and YAGO $\mathcal{KB}$ (version 3.0.2). The architecture of the YAGO system is based on themes. Each theme is a set of facts. A fact is the equivalent of an RDF triple (s,p,o). YAGO has defined the context relation between individuals [30], which we used to extract the set of themes related to our context study which is university domain. The resulting contextual YAGO $\mathcal{KB}$ contains around 5.9×10^6 triples. Note that five (5) sets of triples were generated using LUBM benchmark and Yago knowledge base. The LUBM benchmark provides the Data Generator tool (called UBA) used to create data in the unit of a university. The YAGO knowledge base has more than 10 million entities and contains more than 120 million facts about these entities. It has achieved a precision of 95%. Compatible with RDF, it allows generating data in different formats: Turtle, N-Triple, Literals, etc. *Data-Sets.* five (5) sets of triples were generated using LUBM benchmark and Yago KB. Number of instances (N-Triple format) is shown in Table 4.

Table 4. Dataset characteristics.

Concepts	Ontological instances	$\mathcal{KB}$ Instances
3 Universities	21 057	15988
6 Universities	42 115	96 962
9 Universities	63 173	123 004
12 Universities	84 231	$3,9 \times 10^5$
15 Universities	105 289	$2,5 \times 10^6$

Five (5) sets of ontologies with respectively 3, 6, 9, 12 and 15 universities are generated using Data Generator tool (UBA) provided by the LUBM benchmark. This tool allows creating a domain ontology *University*. Each university consists of 15 to 25 departments. In each department, different categories of *Professors*, *Students*, *GraduateStudents*, *Courses*, etc. can be found. The UBA generates OWL data over the LUBM ontology in the unit of a university. These data are repeatable and customizable, allowing us to specify the seed for random number generation of instances. $\mathcal{KB}$ instances are generated from contextual YAGO $\mathcal{KB}$, five (5) data-sets of instances containing respectively 3, 6, 9, 12 and 15 universities (same number of universities than ontology instances in order to make a comparison). Each university contains different dependencies with the classes: Students, Courses, cities, publications, etc. Number of instances (N-Triple format) used in both approaches is shown in Table 4. However, we have added some contradictory cases to test their influence on the semantic data integration. *(1) Deployment of Data Sources and* $\mathcal{TDW}$: We have created five Oracle *SDBs* using generated data-sets and deployed the $\mathcal{TDW}$ schema using Oracle *SDB*. We chose N-Triple format (.nt) to load instances using Oracle SQL*Loader.

(2) Oracle Database Tuning: $\mathcal{TDW}$ schema was optimized using Btree indexing triples and sparql query hints. Some PL/SQL APIs are also invoked after each load of significant amount of data. The API SEM_PERF.GATHER_STATS Collects stats for sources models and SEM_APIS.ANALYZE_MODEL for $\mathcal{TDW}$ model in the semantic network graph. The memory SGA and PGA are also increased to 2 GB.

(3) Inference Engine: Oracle has incorporated a reasoner engine defined based on *TrOWL* and *Pellet* reasoners. Oracle provides full support for native inference in the database for RDFS, RDFS++, OWLPRIME, OWL2RL, etc. It uses forward chaining to do the inference. It compiles entailment rules directly to SQL and uses Oracle's native cost-based SQL optimizer to choose an efficient execution plan for each rule. The following is an example of user defined rules applied, they are saved as records in tables. $Rule_1$: co-author rule: $authorOf(?A1, ?P) \wedge authorOf(?A2, ?P) \rightarrow coAuthor(?A1, ?A2)$. *(4) Hardware:* Our evaluations were performed on a laptop computer (HP Elite-Book 840 G2) with an Intel(R) CoreTM i5-5200U CPU 2.20 GHZ and 8 GB of RAM and a 500 GB hard disk. We use Windows10 64 bits. Cytoscape[8] is used for visualization. We use Oracle Database 12c release 1 that offers RDF Semantic Graph features of Oracle Spatial and Graph.

Obtained Results. We evaluate our proposal based on the following criteria: *Criterion 1: ETL Algorithm Complexity.* The algorithm is implemented based on semantic ontologies (classes and properties) and graph theory, where nodes represents concepts and instances, edges for roles and labels for definitions. We examine the number of iterations of our algorithm to generate ETL process as flow or graph. In this case, we are interesting on the time complexity. The algorithm is based on concepts searches (Tbox for intentional mappings i.e. mappings only between classes and properties and not between instances). The time

---

[8] http://www.cytoscape.org/.

complexity is $O(n)$, where n represents the number of involved classes or nodes. Figure 10a shows the number of iterations by classes. It indicates a polynomial time. This finding shows the feasibility and efficiency of our approach.

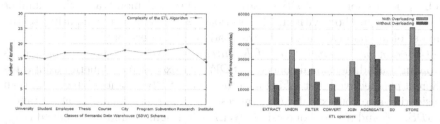

(a) Complexity of the proposed ETL algo-   (b) Evaluation time per ETL operator be-
rithm   fore and after overloading.

Fig. 10. Complexity and evaluation time of the ETL process.

Criterion 2: Evaluation time per ETL operator before and after overloading. We run the ETL Algorithm for both scenarios (without overload for ontology and $\mathcal{KB}$, and with overload for both) to populate the target schema of semantic $\mathcal{TDW}$. We measure the time spent to run each ETL operator. Figure 10b shows the results obtained. Our approach improves the performance time spent by overloaded ETL operator in an 18%. This is due to one call of the functions related to ETL operators done by the compiler, instead of multiple calls in a case without overload.

Criterion 3: Scalability of the proposed solution. The ETL Algorithm populates the target schema of semantic $\mathcal{TDW}$ using an overload of ETL operators. We measure the time spent to integrate data sources having different sizes. Note that time spent to load all instances is equal 3,2 min. Figure 11a illustrates the results obtained where for each triple size loaded using overload approach, corresponding time performance is shown in milliseconds. The result remains reasonable w.r.t. the size of the stored instances. This is proof the scalability of our approach.

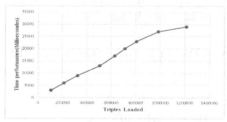
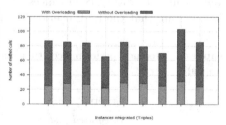
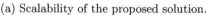

(a) Scalability of the proposed solution.   (b) Number of method calls

Fig. 11. Scalability and number of calls.

Criterion 4: Number of method Calls. We consider a set of $\mathcal{SDB}$ participating in the $\mathcal{TDW}$. We run the ETL Algorithm from two different perspectives: first taking in account the overload of ETL process, second without considering it. Figure 11b shows the number of methods calls with and without overloading of ETL operators for each $\mathcal{SDB}$ integrated. It clearly demonstrates that number of invocation method without the overload is much higher comparing to the number of method calls using the overload of ETL operators.

Criterion 5: Response time of Requirements. We consider a set of queries involving multidimensional concepts of $\mathcal{DW}$. We executed the queries from three different perspectives: first taking in account only $\mathcal{SDB}$ instances integrated, then considering also contextual $\mathcal{KB}$ instances added, and last including all instances of $\mathcal{SDB}$, $\mathcal{KB}$ and inferred one using OWLPrime fragment and user defined rules. Figure 12 shows that query result size is most strongly important using $\mathcal{KB}$ and inferred instances.

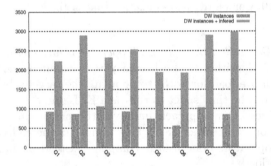

Fig. 12. Query answering before and after inference

Criterion 6: Inference performance. We evaluate inference performance using OWLPrime fragment and user defined rules. First, we define two different models where each one stores sub-graph of $\mathcal{DW}$. The first model stores instances integrated from $\mathcal{SDB}$ sources, the second one stores instances extracted from both $\mathcal{SDB}$ sources and contextual $\mathcal{KB}$. Then, we used reasoner mechanism to infer instances from both models to show number of instances inferred using $\mathcal{SDB}$ sources unified with $\mathcal{KB}$ instances. Table 5 shows results obtained. It clearly demonstrate that the ETL overload manages inference in less time.

Table 5. Inference performance: time and number of instances (triples and RDF graphs).

Criteria	SDB	SDB + $\mathcal{KB}$
Integrated instances	315 865	5.4×10^6
Inferred instances	6k	34K
Time inference (minutes)	4.4	7.2

8   Conclusion

In this paper, we attempt to identify the strong interaction between two important Vs brought by Big Data Era which are Value and Variety. To get more value, data warehouse designers need to consider more sources with a high variety. Another important aspect that has to be considered when building a warehouse is the variety of its system store. Our deep analysis of the most important state of art allows us to identify the ETL landscape that we called it ETL environment. Without this environment, it is hard to deal with variety. A rich discussion and comparison of the most important state of art works related to the unification and genericity efforts of elements of ETL environment are presented. Thanks to Model Driven Engineering techniques, we make generic all elements of the ETL environment, where meta models are proposed for each element. This genericity contributes in overloading all ETL operators in order to reduce their development costs (prototyping) and consequently their performance. Examples of instantiation of three major classes of databases (relational, semantic and graph) are given. Our efforts of genericity facilitate the multi-store deployment. Finally, an evaluation of our proposal to study the effect of overloading on the performance of different operators is also given. A tool available at Youtube is also given.

Currently, we are working the scalability of our proposal using considering a large set of dynamic data sources.

References

1. El Akkaoui, Z., Mazón, J.-N., Vaisman, A., Zimányi, E.: BPMN-based conceptual modeling of ETL processes. In: Cuzzocrea, A., Dayal, U. (eds.) DaWaK 2012. LNCS, vol. 7448, pp. 1–14. Springer, Heidelberg (2012). https://doi.org/10.1007/978-3-642-32584-7_1
2. Ali, S.M.F., Wrembel, R.: From conceptual design to performance optimization of ETL workflows: current state of research and open problems. VLDB J. **26**(6), 777–801 (2017)
3. Baader, F., Calvanese, D., McGuinness, D.L., Nardi, D., Patel-Schneider, P.F. (eds.): The Description Logic Handbook: Theory, Implementation, and Applications. Cambridge University Press, Cambridge (2003)
4. Berkani, N., Bellatreche, L.: A variety-sensitive ETL processes. In: Benslimane, D., Damiani, E., Grosky, W.I., Hameurlain, A., Sheth, A., Wagner, R.R. (eds.) DEXA 2017. LNCS, vol. 10439, pp. 201–216. Springer, Cham (2017). https://doi.org/10.1007/978-3-319-64471-4_17
5. Berkani, N., Bellatreche, L., Khouri, S.: Towards a conceptualization of ETL and physical storage of semantic data warehouses as a service. Cluster Comput. **16**(4), 915–931 (2013)
6. Calvanese, D., De Giacomo, G., Lenzerini, M., Nardi, D., Rosati, R.: Data integration in data warehousing. Int. J. Coop. Inf. Syst. **10**(3), 237–271 (2001)
7. Calvanese, D., Lenzerini, M., Nardi, D.: Description logics for conceptual data modeling. In: Chomicki, J., Saake, G. (eds.) Logics for Databases and Information Systems, vol. 436, pp. 229–263. Springer, Boston (1998). https://doi.org/10.1007/978-1-4615-5643-5_8

8. Craig, I.: The Interpretation of Object-Oriented Programming Languages. Springer, London (2002). https://doi.org/10.1007/978-1-4471-0199-4

9. DeWitt, D.J., et al.: Split query processing in polybase. In: Proceedings of the 2013 ACM SIGMOD International Conference on Management of Data, pp. 1255–1266. ACM (2013)

10. Dong, X.L., Srivastava, D.: Big data integration. PVLDB 6(11), 118 (2013)

11. Duggan, J., et al.: The BigDAWG polystore system. ACM SIGMOD Rec. 44(2), 11–16 (2015)

12. Inmon, W.H.: Building the Data Warehouse. Wiley, Hoboken (2002)

13. Mazón, J.-N., Trujillo, J.: An MDA approach for the development of data warehouses. In: JISBD, pp. 208–208 (2009)

14. Jean, S., Bellatreche, L., Ordonez, C., Fokou, G., Baron, M.: OntoDBench: interactively benchmarking ontology storage in a database. In: Ng, W., Storey, V.C., Trujillo, J.C. (eds.) ER 2013. LNCS, vol. 8217, pp. 499–503. Springer, Heidelberg (2013). https://doi.org/10.1007/978-3-642-41924-9_44

15. Khouri, S., Semassel, K., Bellatreche, L.: Managing data warehouse traceability: a life-cycle driven approach. In: Zdravkovic, J., Kirikova, M., Johannesson, P. (eds.) CAiSE 2015. LNCS, vol. 9097, pp. 199–213. Springer, Cham (2015). https://doi.org/10.1007/978-3-319-19069-3_13

16. Kolev, B., Valduriez, P., Bondiombouy, C., Jiménez-Peris, R., Pau, R., Pereira, J.: CloudMdsQL: querying heterogeneous cloud data stores with a common language. Distrib. Parallel Databases 34(4), 463–503 (2016)

17. Lenzerini, M.: Data integration: a theoretical perspective. In: ACM SIGACT-SIGMOD-SIGART Symposium on Principles of Database Systems, pp. 233–246 (2002)

18. Luján-Mora, S., Vassiliadis, P., Trujillo, J.: Data mapping diagrams for data warehouse design with UML. In: Atzeni, P., Chu, W., Lu, H., Zhou, S., Ling, T.-W. (eds.) ER 2004. LNCS, vol. 3288, pp. 191–204. Springer, Heidelberg (2004). https://doi.org/10.1007/978-3-540-30464-7_16

19. Nakuçi, E., Theodorou, V., Jovanovic, P., Abelló, A.: Bijoux: data generator for evaluating ETL process quality. In: ACM DOLAP, pp. 23–32 (2014)

20. Nebot, V., Berlanga, R.: Building data warehouses with semantic web data. Decis. Support Syst. 52(4), 853–868 (2012)

21. Ong, K.W., Papakonstantinou, Y., Vernoux, R.: The SQL++ unifying semi-structured query language, and an expressiveness benchmark of SQL-on-Hadoop, NoSQL and NewSQL databases. CoRR, abs/1405.3631 (2014)

22. Raventós, R., Olivé, A.: An object-oriented operation-based approach to translation between MOF metaschemas. Data Knowl. Eng. 67(3), 444–462 (2008)

23. Rodriguez, M.A., Neubauer, P.: Constructions from dots and lines. CoRR, abs/1006.2361 (2010)

24. Shmueli, O., Tsur, S.: Logical diagnosis of LDL programs. New Gener. Comput. 9(3/4), 277–304 (1991)

25. Simitsis, A., Vassiliadis, P., Sellis, T.-K.: Optimizing ETL processes in data warehouses. In: ICDE, pp. 564–575 (2005)

26. Simitsis, A., Wilkinson, K., Castellanos, M., Dayal, U.: Optimizing analytic data flows for multiple execution engines. In: Proceedings of the 2012 ACM SIGMOD International Conference on Management of Data, pp. 829–840. ACM (2012)

27. Simitsis, A., Wilkinson, K., Dayal, U., Castellanos, M.: Optimizing ETL workflows for fault-tolerance. In: ICDE, pp. 385–396 (2010)

28. Skoutas, D., Simitsis, A.: Ontology-based conceptual design of ETL processes for both structured and semi-structured data. Int. J. Semant. Web Inf. Syst. **3**(4), 1–24 (2007)

29. Stonebraker, M.: Technical perspective - one size fits all: an idea whose time has come and gone. Commun. ACM **51**(12), 76 (2008)

30. Suchanek, F.M., Kasneci, G., Weikum, G.: Yago: a core of semantic knowledge. In: WWW, pp. 697–706 (2007)

31. Trujillo, J., Luján-Mora, S.: A UML based approach for modeling ETL processes in data warehouses. In: Song, I.-Y., Liddle, S.W., Ling, T.-W., Scheuermann, P. (eds.) ER 2003. LNCS, vol. 2813, pp. 307–320. Springer, Heidelberg (2003). https://doi.org/10.1007/978-3-540-39648-2_25

32. Tziovara, P., Vassiliadis, P., Simitsis, A.: Deciding the physical implementation of ETL workflows. In: DOLAP, pp. 49–56 (2007)

33. Vassiliadis, P.: A survey of extract-transform-load technology. IJDWM **5**(3), 1–27 (2009)

34. Vassiliadis, P., Simitsis, A., Baikousi, E.: A taxonomy of ETL activities. In: ACM DOLAP, pp. 25–32 (2009)

35. Vassiliadis, P., Simitsis, A., Georgantas, P., Terrovitis, M., Skiadopoulos, S.: A generic and customizable framework for the design of etl scenarios. Inf. Syst. **30**(7), 492–525 (2005)

36. Vassiliadis, P., Simitsis, A., Skiadopoulos, S.: Conceptual modeling for ETL processes. In: DOLAP, pp. 14–21 (2002)

37. Vassiliadis, P., Simitsis, A., Skiadopoulos, S.: Modeling ETL activities as graphs. In: DMDW, pp. 52–61 (2002)

38. Wilkinson, K., Simitsis, A., Castellanos, M., Dayal, U.: Leveraging business process models for ETL design. In: Parsons, J., Saeki, M., Shoval, P., Woo, C., Wand, Y. (eds.) ER 2010. LNCS, vol. 6412, pp. 15–30. Springer, Heidelberg (2010). https://doi.org/10.1007/978-3-642-16373-9_2

39. Zhu, M., Risch, T.: Querying combined cloud-based and relational databases. In: 2011 International Conference on Cloud and Service Computing (CSC), pp. 330–335. IEEE (2011)

eVM: An Event Virtual Machine Framework

Elio Mansour[1]([✉]), Richard Chbeir[2], and Philippe Arnould[1]

[1] Univ Pau & Pays Adour/E2S-UPPA, LIUPPA, EA3000, Mont-de-Marsan, France
{elio.mansour,philippe.arnould}@univ-pau.fr
[2] Univ Pau & Pays Adour/E2S-UPPA, LIUPPA, EA3000, Anglet, France
richard.chbeir@univ-pau.fr

Abstract. Information and communication technology (ICT) is impacting our daily lives more than ever before. Many existing applications guide users in their daily activities (e.g., navigation through traffic, health monitoring, managing home comfort, socializing with others). Although these applications are different in terms of purpose and application domain, they all detect events and propose actions and decision making aid to users. However, there is no usage of a common backbone for event detection that can be instantiated, re-used, and reconfigured in different use cases. In this paper, we propose eVM, a generic event Virtual Machine able to detect events in different contexts while allowing domain experts to model and define the targeted events prior to detection. eVM simultaneously considers the various features of the defined events (e.g., temporal, geographical), and uses the latter to detect different feature-centric events (e.g., time-centric, location-centric). eVM is based on different components (an event query language, a query compiler, an event detection core, etc.), but mainly the event detection modules are detailed here. We show that eVM is re-usable in different contexts and that the performance of our prototype is quasi-linear in most cases. Our experimental results showed that the detection accuracy is improved when, besides spatio-temporal information, other features are considered.

Keywords: Event detection · Semantic clustering
Formal concept analysis.

1 Introduction

During the last decades, the world has witnessed important changes. Most notably, information and communication technology (ICT) has become a major part of our lives. Recent research efforts have brought forward massive advances in the field of data management, thus easing the way for ICT to be tightly integrated in various application domains (e.g., social, medical, home/building management, energy management, navigation systems, industry and manufacturing). As a result, we now have numerous applications that help users in their everyday tasks. To give a few examples, consider social applications that allow users to socialize, comment on social issues and gatherings, and share multimedia data

© Springer-Verlag GmbH Germany, part of Springer Nature 2018
A. Hameurlain et al. (Eds.): TLDKS XXXIX, LNCS 11310, pp. 130–168, 2018.
https://doi.org/10.1007/978-3-662-58415-6_5

(e.g., photos, videos) taken by different users, during life events [27] through a simple interface. Social media have evolved greatly over the past decade. As a result, more users are now connected to collaborative and information sharing platforms (e.g., Facebook, Google+) and other social sharing applications (e.g., Iphotos[1]). The medical domain is also witnessing the evolution of ICT. Nowadays, patients use personal sensors to monitor their health. Several studies [17,20,41] detail the usage of wearable sensors and data mining techniques in order to monitor a patient's heart condition, or gait analysis (i.e., a person's manner of walking). Buildings and homes are also evolving. With the integration of sensor networks and advanced data processing techniques, we are currently in the era of automated homes and smart buildings [6,43]. In contrast with traditional buildings/homes, these dynamic systems provide a more comfortable [22], sophisticated [40], and healthy [46] environment for occupants. Other works [1] focus more on energy management in these infrastructures. In this regard, the integration of data mining and sensor networks reduced energy costs for building/home owners, and more importantly diminished the environmental impact of the aforementioned infrastructures. ICT is also impacting the way we travel, drive, and move around in our cities. Navigation and driving guidance systems [44] help users on the roads everyday. Using geo-localization, sensor networks, and modern telecommunication, drivers can now avoid traffic jams/congestion, and report occurring incidents. This is making roads safer and more accessible for everyone. The integration of ICT even reached the industrial domain [23]. Currently, factory managers are using machine monitoring applications that allow them to better schedule machine maintenance, thus preventing faults and breakdowns.

All of the aforementioned works are different (in terms of application domains, purposes, and objectives). However, they share the need to detect important events. To achieve this, many works [8,9,32,36,37] have evolved around the organization of shared data on platforms and applications using clustering techniques. Some social-based studies [8,32,35,37] are based on metadata (e.g., Facebook organizes multimedia content based on publishing timestamps, Iphotos combines photo creation timestamps and locations to organize a user's photo library). Others [26,28,31] use the visual attributes of shared objects (e.g., textures, colors) coupled with metadata in order to detect several events. Medical/health monitoring applications [17,20,41] detect health-related events for a specific patient. In the case of urgent/life threatening events, event detection allows an efficient and timely intervention from the patient's doctor. Otherwise, information regarding the detected events can be used for deep analysis of the medical issue, testing, diagnosis, and better decision making. Home/building and energy management works [1,22,40,46] also detect events. For example, detecting if a person entered a room in order to automatically turn on the lights, or detecting if the building is empty to turn off the heating and reduce energy consumption. Furthermore, driving guidance systems [44] need to detect events such as accidents, traffic jams, and road maintenance in order to steer drivers away from these events'locations. Finally, machinery monitoring applications [23]

---

[1] http://www.apple.com/ios/photos.

automatically detect events that could lead to machine breakdowns and faults in order to optimize maintenance scheduling.

All the aforementioned works share common principles: (i) they need to detect a set of events; and (ii) they need to extract data and apply at least one data mining technique (e.g., clustering, classification). The main differences are two fold: (1) the targeted events and therefore their features, and (2) the choice of data mining/event detection technique.

Even-though all these works detect events, there is no usage of a common backbone for event detection that can be instantiated in different contexts/application domains. This is restrictive and costly since existing solutions suffer of the following issues: (i) the absence of an evolutionary approach capable of coping with needs that change over time; (ii) the absence of extensibility regarding the integration of new plug ins/complementary systems to an existing event detection approach; (iii) the difficulty of integrating different event-related modifications in the development; (iv) the impossibility of reusing the same framework to detect other events in various domains/contexts; and (v) the lack of expert input, i.e., providing a module where one can provide his own input on how to define the corresponding events prior to detection. In addition, sometimes data shared/published is heterogeneous, and this generates some technical aspects to be considered: multi-modality (the ability to consider various features and datatypes at once in the processing), incremental processing (allowing a continuous integration of new data (e.g., new sensor data, new photos shared/published on different dates/times) in the set of already processed data), and multi-source processing (considering various data providers at once). Thus, there is a need to design a generic event detection approach, considering expert input, and event features, to provide a more reusable approach.

To answer this need, we propose eVM: a re-usable architecture for automatic and generic detection of "feature-centric" events. Our framework is composed of several main components: (1) An event Query Language (eQL): one can use the eQL to create, update, delete, select, or insert datatypes and event features, thus defining the targeted events. In addition, one can define several key event features. Based on the latter, our approach detects the corresponding feature-centric events. This allows the framework to be generic and re-usable in different event detection contexts/domains while allowing domain experts to provide their input; (2) An easy to integrate API containing an Event Query Compiler for eQL query processing and an Event Detection core. This makes eVM evolutionary, extensible, and easy to integrate with other modules/systems and programing languages; and (3) a storage space where various repositories are available for the storage of various event related data (e.g., data objects, event definitions, detected feature-centric events).

eVM's clustering technique simultaneously considers the various features, objects shared by different sources (several data producers) on different dates/-times, as well as data from one source (i.e., to organize a source's data library). eVM is based on an adaptation of FCA (Formal Concept Analysis) [16,42], a backbone that provides a multi-modal, incremental, and multi-source clustering technique that handles high dimensional data, and requires low human intervention.

In order to validate our approach, we implemented eVM as a cross-platform mobile application in order to evaluate the approach in different real case scenarios. Our experimental results, on ReSEED Dataset [34] and another simulated one, show that the event detection accuracy is improved when additional features (i.e., other than time and geo-location) are taken into consideration. In addition, our performance results show quasi-linear behavior in most cases.

The rest of the paper is organized as follows. Section 2 reviews event detection works. Section 3 introduces FCA and defines the eVM approach. The implementation and evaluation are discussed in Sect. 4. Section 5 reviews clustering techniques. Finally, Sect. 6 concludes and highlights future perspectives.

2   Related Work

In this section, we review event definitions, types, and features before detailing event detection works in different areas (e.g., social event detection, medical event detection, sensor event detection). Since, in most cases, events are detected based on incoming raw data, without having any prior knowledge on the occurring events, main approaches [8,28,31,32,37] use unsupervised clustering techniques. Since there are no commonly adopted criteria, we propose the following set of criteria to compare the referenced works:

1. *Re-Usability*: This criterion examines the possibility (*yes* or *no*) of using the same approach as an event detection backbone for different targeted events in different contexts/domains (cf. Sect. 1 - limitation (iv)).
2. *Domain Specific Expertise*: This criterion measures the possibility (*yes* or *no*) of taking into account input from domain experts in the event definition process (cf. Sect. 1 - limitation (v)).
3. *Evolution*: This criterion examines (*yes* or *no*) if an event detection approach can evolve and adapt to the changing event definition/detection needs over time (cf. Sect. 1 - limitation (i)).
4. *Extensibility*: This criterion measures (*yes* or *no*) the capability of integrating new plug-ins/external modules in an existent event detection approach (cf. Sect. 1 - limitation (ii)).
5. *Ease of Integration*: This criterion denotes (*yes* or *no*) an approach's capability of integrating event-related modifications in the development (cf. Sect. 1 - limitation (iii)).

In addition to the aforementioned criteria, we also consider other technical requirements/criteria such as:

6. *Multi-modality*: This criterion states (*yes* or *no*) if multiple event features having different datatypes are considered (e.g., *social, topics* (textual information), *sensor-related* (scalar and multimedia information)) in addition to time (datetime) and locations (GPS coordinates or textual location description) for improved event detection.

7. *Multi-source*: This criterion indicates (*yes* or *no*) if multiple data sources (various publishers, data producers) are considered. This is important since various sources can provide valuable event related data.
8. *Incremental (continuous) processing*: This criterion considers the possibility (*yes* or *no*) of processing incoming data without having to repeat the entire processing, because data producers could share event related data on different dates/times.
9. *Level of human intervention*: This criterion measures (*high, moderate,* or *low*) how frequently users participate in the event detection process; since huge amounts of data are shared, it is important that user interventions become less frequent; we consider low intervention if users provide data input and initial configuration; moderate if users also intervene in result correction/optimization; and high intervention when users participate in the whole process.

In the following, we begin by defining events, before detailing research works from various domains in which event detection has had noticeable impact (e.g., social event detection, sensor event detection).

2.1   Basic Definition of an Event

In the literature, many works [2,3] define events as a happening that takes place at a particular time and location. Thus, emphasizing the importance of two main event features: (i) temporal; and (ii) spatial. All events are associated with these two features, since they answer the most common inquiries i.e., where and when. Nonetheless, additional event features are useful to describe the context and semantics of an event (e.g., social, political, medical). The additional features differ from an event to another.

Events are categorized into different types, regardless of their contexts: (i) atomic or primitive events are the simplest events that can occur in a system. They cannot be decomposed into any smaller entity; (ii) composite or complex events are high level derived events, and are defined by combining constituent events. The latter can be atomic, or/and composite [2]. In our proposal, we currently consider atomic events, nonetheless the framework is re-usable and extensible, and can easily integrate a module for event composition that allows the detection of composite events.

2.2   Event Detection Applications

As mentioned before, event detection covers a large spectrum of application domains. From social event detection and sharing applications to environmental monitoring (e.g., detecting fire hazards in forests, level of air pollution in a city), energy management (e.g., detecting energy wastes in smart buildings), industrial processes (e.g., detecting events that disrupt production flow in a factory, detecting faults and machine maintenance issues), and medical event detection (e.g., monitoring a patient's heart condition) in various sensor networks. In all the aforementioned works, event detection requires a sensing (data collection) phase. During this phase, data is collected either through social means (shared images, videos, and posts on social media platforms) or physical equipments (e.g., sensor observations produced by deployed sensors). Therefore, we discuss event detection works by their respective categories: (i) Social-based; and (ii) Equipment-based works.

Social-Based Works: In the literature, several Social Event Detection (SED) approaches have emerged for detecting events. SED approaches can be grouped into two categories: approaches that rely on the metadata of shared objects (e.g., photos, videos, tweets) [8,32,35,37], denoted metadata-based, and approaches that rely on visual attributes (e.g., colors, shapes) and metadata [13,14,26,28, 31], denoted hybrid.

- Metadata-based approaches: In [32], the authors aim to detect social events based on image metadata, using temporal, geo-location, and photo-creator information. They perform a multi-level clustering for these features. In [8], the authors use time and GPS data to cluster photos into events using the mean-shift algorithm. First, the authors find baseline clusters based on time, then GPS location attributes are integrated. In [35], the authors use time and location information from twitter feeds to detect various events (e.g., earthquakes). In [37], the authors rely on textual tags such as time, geo-location, image title, descriptions, and user supplied tags to cluster photos into events, thus detecting soccer matches that happened in Madrid. These approaches need moderate human intervention. However, they are not incremental nor extensible. Metadata is also used by stand-alone applications for photo management to detect social events in a user's multimedia library. These applications require no human intervention, they automatically cluster objects found in a library. However, they do not consider other event features (e.g., social, topic), nor other photo sources (photos taken by other participants/collaborators). They mainly focus on time and location, photos taken at the same day and place of an event are merged with the event.
- Hybrid approaches: Many hybrid approaches rely on both visual and metadata attributes. In [28] and [31], the authors combine visual object attributes with temporal information, geo-locations, and user-supplied tags for their clustering procedures. Visual and tag similarity graphs are combined in [28] for the clustering. While in [31] the authors divide the geographical map of the world into square tiles and then extract the photos of each tile using geo-location metadata. They later use other metadata combined with visual features to detect objects and events. In [14,26], the authors combine temporal metadata with visual attributes for annotation and event clustering purposes. In [13], the author relies more on temporal metadata than visual attributes for correct event detection, since he considers that photos/videos associated with one event are often not visually similar. Hybrid approaches consider different types of object attributes (e.g., visual, temporal, geo-locations). However, regrouping visually similar objects does not imply that they belong to the same event. Therefore, metadata is required to boost the accuracy of such

approaches. Since these methods process visual attributes (e.g., through photo/video processing techniques), they end up having a higher processing cost than the approaches that only process metadata. Some approaches require more human intervention, because they prompt the user to correct/optimize the results.

Equipment-Based Works: With the evolution of sensor technology [11], sensor networks have been highly used for various purposes (e.g., personal sensing, social sensing, environmental monitoring), especially since the sensor data modeling aspects have been widely covered recently (e.g., through ontologies such as the Semantic Sensor Network SSN [12]). Sensors produce observations related to certain properties (e.g., temperature, movement, humidity). Regarding event detection in sensor networks, events are usually composed of sensor features (e.g., temperature) alongside spatio-temporal features since every sensor observation is mapped to an instant in time and a specific location. Many works [2, 25] agree that sensor observations are considered atomic events (e.g., temperature rise event), therefore works regarding sensor data fusion [4] could target the composite events. In the following, we detail some event detection works in sensor networks, based on the application domains.

- Medical Event Detection: In [17], the authors use lightweight, wearable, and durable sensors to monitor patients' gait (i.e., the manner of walking). People who suffer from strokes or spinal cord injuries, tend to have abnormal gaits. During medical treatment, it is beneficial to detect gait events when they occur (e.g., initial foot contact). The authors propose two different ways for detecting such events, one using accelerometer data, and another using foot switch data (i.e., data from pressure/force sensor). In both cases, the event features are spatio-temporal, and sensor-related (i.e., accelerometer, pressure, force). The authors test both cases on normal, slow, and altered walking subjects and achieve near real time accurate detection of abnormal gait events. In [20], the authors declare that a variety of measurements are required for gait analysis (e.g., stride, step lengths, cadence, gait velocity). In order to acquire such measurements, the system needs to know when and where each foot leaves and touches the ground again. Therefore, the authors take interest in detecting the following events: (i) foot end contact (EC); and (ii) initial contact (IC). Thus, the authors propose an approach for IC and EC event detection using linear accelerometers and angular velocity transducers. They then use the event detection results to analyze gait patterns of healthy and injured individuals. In [41], the authors propose a wireless smart sensor for heart monitoring. The aim is to detect life threatening events such as cardiac arrhythmia for patients with heart related issues. The sensor monitors heart rate and ECG signals to detect the aforementioned events in real time.
- Environmental Event Detection: In environmental monitoring scenarios, the sensor network contains more nodes (compared to personal medical sensing), and the spatio-temporal data acquisition intervals are wider. For example, to detect high air pollution events in a city, a huge number of air quality sensors

should be deployed. In [15], the authors detect wildfire events in the wild by collecting sensor data such as temperature, relative humidity, and barometric pressure. In addition, they integrate spatial features by using a GPS unit in order to localize the detected events. Information is communicated using a wireless sensor network. In [45], the authors propose an approach for real-time forest fire detection. They rely on spatio-temporal information and fire event context features such as relative humidity, temperature, smoke, and wind speed. They produce a report of abnormal atomic events (e.g., high temperature, smoke rising), and a real time forest fire danger rate from the collected data. Then they use a neural network to detect the fire events. In [24], the authors use crowd-sensing in order to detect and monitor air quality related events in a city. In addition to time and geo-localization, air quality events share features that are related to the context of air quality (e.g., carbon monoxide (CO), air pressure, nitrogen dioxide (NO2), and temperature). They also develop an android mobile phone application to display results to end users.

Table 1. Event detection works comparison

Criterion	Event detection category				
	Social-based works			Equipment-based works	
	Metadata		Hybrid	Medical	Environmental
	$[8,32,35,37]$	Stand-alone applications	$[13,14,26,28,31]$	$[17,20,41]$	$[15,24,45]$
Re-usability	No	No	No	No	No
Domain specific expertise	No	No	No	No	No
Evolution	No	Partially[a]	No	No	No
Extensibility	No	Partially[a]	No	No	No
Ease of integration	No	No	No	No	No
Multi-modality	No	Yes	Partially[a]	Yes	Yes
Multi-source	Partially[a]	Partially[a]	Partially[a]	No	No
Incremental processing	No	Partially[a]	No	Yes	Yes
Level of human intervention	Low - moderate	Low	Moderate - high	Moderate	Moderate

[a] Partially states that not all approaches of a category comply with the criterion.

Discussion: Table 1 summarizes the evaluation of event detection approaches based on the aforementioned criteria. Concerning Social-based works, metadata-based approaches [8,32,37] need low to moderate human intervention and provide good event detection accuracy, since metadata describes data related to the events (e.g., dates, locations, tags). However, these works lack the incremental processing needed to match the flow of publishing/sharing. Recently, incremental processing was integrated in some works (e.g., iPhotos). Hybrid methods [13,14,26,28,31] are costly computation-wise and require human intervention thus making continuous processing hard to implement [29]. In contrast, these methods offer more event features by combining visual attributes with metadata to improve accuracy. Finally, the two categories of works are not re-usable in different contexts nor fully consider the various features and modalities (datatypes) in event detection. They also do not consider domain expert input when defining the targeted events. Concerning Equipment-based works, we split these approaches based on their application domains. Nonetheless, they share two common characteristics: (i) they rely on a data acquisition networks (constituted of one or more sensors); and (ii) although they target different events, spatio-temporal event features are used by all methods. When considering the latter features, the chosen granularity can vary based on the application (e.g., for the spatial feature: we consider cities for environmental monitoring and the specific indoor location of a patient in a fall detection system). What differentiates the event definition from one approach to another are the specific (context related) features (e.g., temperature, movement, humidity). No current approach allows domain experts to contribute in event definition, i.e., the same event (e.g., abnormal gait) has variant definitions in different approaches. These works are not re-usable in different domains and contexts. When considering the other criteria we find that the level of human intervention varies from an approach to another. Finally even though these approaches are incremental, in most cases they are not extensible.

3   The Event Virtual Machine (eVM) Approach

In this section, we detail the eVM framework. Firstly, we provide an overview of our proposal. Then, we focus more on the clustering technique used in our event detection process. Finally, we detail our query language, modules and algorithm.

3.1   Approach Overview

As previously discussed in Sect. 2.1, existing approaches heavily rely on time and location when defining the targeted events. In order to tackle specific events, additional context related information (e.g., social, medical) are added. Nonetheless, existing works, such as [17,32,35], provide a static event definition that does not allow modifications (adding, removing event features). Since event definition and detection needs change over time, or from one context/domain to another, we provide a generic and extensible event definition in eVM. We model events as $2D^+$ spaces (having at least two dimensions: temporal and spatial). Depending on the targeted events, one can specify/create additional dimensions, representing the contextual event features such as *topics* (based on *tags* or *annotations*), *temperatures* (based on *sensor readings*) and various levels of granularities for each feature (e.g., year-month-week-day-hour for time, country-region-city-street for location). Micro granularities can also be considered depending on the application purpose (e.g., time: minutes-seconds for short events, location: human body-part of a body for medical applications). This way of defining events is dynamic, allows domain expert input, and is extensible. Nonetheless, it does not consider yet digital events, where the location dimension should be handled differently (e.g., server attacks that could happen on multiple nodes in the same time.).

Depending on the event detection needs, domain experts can create, insert, update and delete datatypes, event features, granularities, and event definitions using our proposed event Query Language (eQL). Every time a eQL query is submitted, a common Event Query Compiler (cf. Fig. 1) executes the submitted query, thus creating an instance of event detection. Finally, using the defined instance, a common Event Detector mechanism is triggered. This process queries event related data, stored in specific repositories (found in the storage space), and starts detecting events based on the provided event model (definition). The Event Detector integrates FCA as the backbone clustering technique, to provide a multi-modal, dynamic, and incremental approach (the API components are later detailed in Fig. 3). This makes the eVM framework re-usable in different contexts, evolutionary, and easy to integrate with any API friendly programing language. In this paper, we only detail the Event Detection part of this framework.

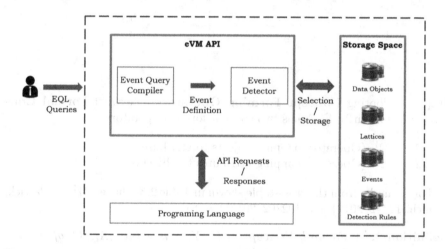

Fig. 1. eVM overview

3.2   FCA Preliminaries and Definitions

After studying various clustering techniques [5,7,19,33], we chose Formal Concept Analysis (FCA) [16,42] as the backbone for the event detection process of our approach. FCA is incremental and multi-modal (criteria 8 and 6). It examines data through object/attribute relationships, extracts formal concepts and orders the latter hierarchically in a Concept Lattice which is generated through a four step process [10]:

Step 1: Defining a **Formal Context** (Definition 1) from the input data, based on object/attribute relations represented in a cross-table.

Definition 1. *A **Formal Context**: is a triplet $\langle X,\ Y,\ I \rangle$ where:*

- *X is a non-empty set of objects*
- *Y is a non-empty set of attributes*
- *I is a binary relation between X and Y mapping objects from X to attributes from Y, i.e., $I \subseteq X \times Y$.* ∎

Table 2 shows an example, where photos are objects and photo attributes (locations, photo creator names, and dates) are attributes. The cross-joins represent the mapping of photos to their respective photo attributes, e.g., photo 1 *was taken in Biarritz by John on 17/08/2016.*

Table 2. Formal context example

	Names				Locations			Dates		
	John	Patrick	Dana	Ellen	Biarritz	Munich	Paris	17/08/2016	12/12/2012	02/02/2016
Photos 1	x				x			x		
2	x				x			x		
3		x				x			x	
4			x			x			x	
5				x			x			x

Step 2: Adopting **Concept Forming Operators** to extract **Formal Concepts** (Definition 2). FCA has two concept forming operators:

- $\uparrow: 2^X \rightarrow 2^Y$ (Operator mapping objects to attributes)
- $\downarrow: 2^Y \rightarrow 2^X$ (Operator mapping attributes to objects).

For example, from the cross-table shown in Table 2, we have $\{3\}^\uparrow = \{$Patrick, Munich, 12/12/2012$\}$ and $\{02/02/2016\}^\downarrow = \{5\}$.

Definition 2. *A **Formal Concept** in $\langle X, Y, I \rangle$ is a pair $\langle A_i, B_i \rangle$ of $A_i \subseteq X$ and $B_i \subseteq Y$ such that: $A_i^\uparrow = B_i \wedge B_i^\downarrow = A_i$.* ∎

Consider the set of photos $A_1 = \{1, 2\}$ and the set of attributes $B_1 = \{$John, Biarritz, 17/08/2016$\}$. $A_1^\uparrow = \{$John, Biarritz, 17/08/2016$\}$ and $B_1^\downarrow = \{1, 2\}$. Thus, since $A_1^\uparrow = B_1$ and $B_1^\downarrow = A_1$, the pair $\langle A_1, B_1 \rangle$ is a Formal Concept.

Step 3: Extracting a **Subconcept/Superconcept Ordering** relation for **Formal Concept** (cf. Definition 2) ordering by defining the most general concept and the most specific concept for each pair. The ordering relation is denoted $\leq$.

For example, from Table 2, let $A_1 = \{3\}$, $B_1 = \{$Patrick, Munich, 12/12/2012$\}$, $A_2 = \{3, 4\}$, and $B_2 = \{$Munich, 12/12/2012$\}$. According to Definition 2, $\langle A_1, B_1 \rangle$ and $\langle A_2, B_2 \rangle$ are formal concepts. In addition, $A_1 \subseteq A_2$ therefore, $\langle A_1, B_1 \rangle \leqslant \langle A_2, B_2 \rangle$. This means that formal concept $\langle A_1, B_1 \rangle$ is a subconcept of formal concept $\langle A_2, B_2 \rangle$ (which is the superconcept).

Step 4: Generating the **Concept Lattice**, which represents the concepts from the most general one (top) to the most specific (bottom). The lattice is defined as the ordered set of all formal concepts extracted from the data (based on $\leq$). A **Concept Lattice** denoted by $\beta(X, Y, I)_{\leqslant}$ is the set of all formal concepts of $\langle X, Y, I \rangle$ ordered by the subconcept/superconcept ordering relation $\leqslant$, i.e.,

$$\beta(X, Y, I) = \{\langle A, B \rangle \in 2^X \times 2^Y | A^{\uparrow} = B, B^{\downarrow} = A\}$$

$\beta(X, Y, I)_{\leqslant}$ associated with a subconcept/superconcept ordering relation is called a concept (Galois) lattice. ∎

For the example shown in Table 2, Fig. 2 illustrates the Concept Lattice. The top node is the concept regrouping all objects having no attributes in common. As we go down in the hierarchy, we notice that concepts have less objects and more shared attributes (a logic OR and AND are applied to objects and attributes respectively when scrolling down towards the bottom node). The bottom node is the most specific, thus regrouping all attributes having zero objects in common. The next section formally describes our eVM framework and how these FCA four steps are integrated and adapted for the clustering of data objects.

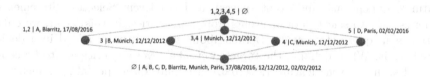

Fig. 2. The concept/galois lattice

3.3 The eVM Framework

In order to organize a set of event-related data objects according to feature-centric events, the eVM's API mechanism is split into four main steps: (i) Event definition & data pre-processing (executed by the Event Query Parser, Event Query Executor, and Pre-Processor modules); (ii) Attribute extraction (executed by the Attribute Extractor module); (iii) lattice construction (executed by the Event Candidates Lattice Builder module); and (iv) event detection (carried out by the Feature-Centric Event Detector and Rule Selector modules). In the following, we detail each processing step and module.

Event Definition and Data Pre-processing: Through the Event Query Compiler, one uses the SQL-like event Query Language (eQL) to define the targeted events. The parser checks the syntax of the submitted queries. Then, the query statements are executed using the Event Query Executor module.

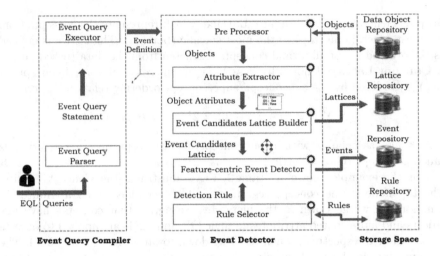

Fig. 3. eVM API components

Query statements are categorized into two groups: (i) Event Definition Statements (queries used to CREATE, UPDATE, INSERT, and DELETE datatypes, features, dimensions, and event spaces); and (ii) Event Selection Statements (queries used to select data or events from repositories. In this paper, we only detail main query statements. Details about eQL will be provided in a dedicated study. When defining the targeted events, several queries are to be considered such as event feature (FEATURE), attribute datatype (ADT), dimension (DIMENSION), and event space (eSPACE) creation. The following syntax shows how to do that:

This query creates an event feature:

```
CREATE FEATURE <feature_id> (
    [label =] <value>,
    [G= ] {<value>},
    [interval =] '1' | '0',
    [gran =]  <function>,
);
```

Where `feature_id` is the unique identifier of the feature, `label` is the feature's label, `G` is a set of granularities associated with the feature, `interval` is a boolean indicating if the feature is generated as an interval (true), or not (false), and `gran` is a function that converts any granularity value to another (related to the same feature). For example, the following describes five (fire) event features each having a label, a set of granularities, an indication about interval construction, and a conversion function:

– $Time_f : \langle 'Time', \{Year, Month, Week, Day, Hour\}, 1, Convert_{Time} \rangle$

- $Geo_f : \langle 'Geo', \{Country, Region, City, Street\}, 0, Convert_{Geo}\rangle$
- $Temp_f : \langle 'Temp', \{value, setofvalues, mean, max, min\}, 1, Convert_{Temp}\rangle$
- $CO2_f : \langle 'CO'_2, \{value, setofvalues, mean, max, min\}, 1, Convert_{CO2}\rangle$
- $Smoke_f : \langle 'Smoke', \{singlevalue, setofvalues\}, 0, Convert_{Smoke}\rangle$

For example, Time, Geo, Temp (temperature), CO2, and Smoke features could be used to define a fire event that can be detected from sensor data in a sensor network.

To create an attribute datatype, the syntax of eQL is as follows:

```
CREATE ADT <adt_id> (
[label =] <value>,
[t =] 'Integer'|'Float'|'Boolean'|'Date'|'Time'|
'Date Time'|'Character'|'String',
[range =] {<value>},
[dist =] <function>,
[f =] <value>,
);
```

Where `adt_id` is the unique identifier of the attribute datatype, `label` is the attribute datatype's label, `t` denotes the primitive data type of the attribute, `range` is the domain of the attribute values, `dist` is the function that returns the distance between any two values of the same attribute datatype, and `f` is the event feature mapped to the attribute data type. To continue with the fire event example, the following describes five attribute data types each having a label, a primitive datatype, a range, a distance function (e.g., time difference for temporal attributes, spatial distance for geographical attributes, temperature, CO2, and Smoke differences between various sensor readings for instance), and an associated event feature:

- $Time_{adt} : \langle 'TimeAttribute', Date, Any, TimeDifference, Time_{feature}\rangle$
- $Geo_{adt} : \langle 'GeoAttribute', String, Any, SpatialDistance, Geo_{feature}\rangle$
- $Temp_{adt} : \langle 'TempAttribute', Float, Any, TempDifference, Temp_f\rangle$
- $CO2_{adt} : \langle 'CO2Attribute', Float, Any, CO2Difference, CO2_f\rangle$
- $Smoke_{adt} : \langle 'SmokeAttribute', Boolean, Any, SmokeDifference, Smoke_f\rangle$

An event is formally defined as a n-dimensional space, denoted event space, where each dimension represents an event feature (e.g., time, social, topic, temperature). To create a dimension, one needs to use the following syntax:

```
CREATE DIMENSION <dimension_id> (
    [o =] <value>,
    [datatype =]  <value>,
);
```

Where `dimension_id` is the unique identifier of the dimension, o is the origin point of a dimension, specifying the first value on `DIMENSION`, and `datatype` denotes the attribute datatype shared by all values on `DIMENSION`, `datatype` $\in$ `ADT` (the set of all attribute datatypes). Since each attribute datatype is mapped to a feature, each dimension is also mapped to/represents a specific event feature. For example, the following describes five event space dimensions, each having an identifier, an origin value, and an attribute data type (and therefore an associated event feature). These event dimensions help define the event space of a fire event:

- $Time : \langle 1, 30/12/2017\ 1{:}30\ pm, Time_{adt} \rangle$
- $Geo : \langle 2, Paris, Geo_{adt} \rangle$
- $Temp : \langle 3, 20\ (degrees\ Celsius), Temp_{adt} \rangle$
- $CO2 : \langle 4, 250 PPM (PartsPerMillion), CO2_{adt} \rangle$
- $Smoke : \langle 5, No, Smoke_{adt} \rangle$

And finally, the following query creates an event space:

```
CREATE eSPACE <eventSpace_id> (
    [D =] {<value>},
    [SO =] {<value>},
);
```

Where `eventSpace_id` is the unique identifier of the event space, D is a set of dimensions that constitute the space (such as D at least contains two dimensions: temporal and spatial), and SO is the set of data objects that belong to the event space (the list of objects is empty at the space creation, after the detection process all data objects are inserted into their respective spaces).

Since all events are heavily mapped to a certain time period and a geo-location, we ensure that the event space is at least two dimensional (i.e., at least the temporal and spatial dimensions exist). The additional dimensions that define the event represent the features that the expert chose. This way the framework is re-usable in different contexts (criterion 1 cf. Sect. 2) and the user is able to customize the event definition in order to get more interesting results (criterion 2 cf. Sect. 2). In addition, he/she specifies the feature (or set of features) that he/she would like to consider as key for the feature-centric event detection.

For example, $eSpace : \langle 1, (Time, Geo, Temp, CO2, Smoke), SO \rangle$ defines a fire event where:

- 1 is the id of $eSpace$
- The five dimensions that define the fire event are:
  - $Time : \langle 1, 30/12/2017\ 1{:}30\ pm, Time_{adt} \rangle$
  - $Geo : \langle 2, Paris, Geo_{adt} \rangle$
  - $Temp : \langle 3, 10\ (degrees\ Celsius), Temp_{adt} \rangle$
  - $CO2 : \langle 4, 250(PPM : PartsPerMillion), CO2_{adt} \rangle$
  - $Smoke : \langle 5, No, Smoke_{adt} \rangle$

- SO is the set of data objects sensed/shared during the fire event, *forall so* ∈ *SO*, *so* has five coordinates (temporal, spatial, temperature, CO2, and Smoke).

A domain expert can also execute delete and update related queries:

```
DELETE [FEATURE|ADT|DIMENSION|eSPACE] <id> (
WHERE Condition,
);

UPDATE [FEATURE|ADT|DIMENSION|eSPACE] <id> (
SET [field_name =] <new value>,
WHERE Condition,
);
```

Once the event definition is established, the Pre-processor module requests event related data from the storage unit (i.e., from the object repository). The purpose of this step is to analyze the attributes of each data object (Definition 3). An attribute is defined as a value associated with an attribute data type. We define a data type function denoted dt, that returns the attribute data type of a value based on the data object attributes.

Definition 3. *A **Data Object** is defined as a 2-tuple, so :* $\langle id, V \rangle$, *where:*

- *id is the unique identifier of a data object*
- *V is a set of attribute values according to a given ADT, such that* $\forall a_i \in ADT \quad \exists v_i \in V \mid dt(v_i) = a_i.$ ∎

The *Pre-processor* will extract objects having attributes related to the features found in the event space. For instance, if one targets sports events having temporal, spatial, and a (sport) topic features, the *Pre-processor* extracts from the data object repository (cf. Fig. 3) all objects having the following attributes: (i) a temporal; (ii) a geo-location; and (iii) a sports-related tag/annotation. Finally, the selected data objects are sent to the Attribute Extractor module.

Attribute Extraction: In this step, the event definition, provided by the incoming event space (*eSpace*), is essential for knowing which data object attributes should be extracted and included in the rest of the processing. The attribute extraction objective is to examine the dimensions that constitute *eSpace* and select the list of attribute data types needed for event detection. The Attribute Extractor module (cf. Fig. 4) initiates a cleaning process via the Converter sub-module in order to have the same units for data object attributes (e.g., having all temperature values in Celsius). The cleaned data objects are stored in the data objects repository (cf. Fig. 3). Finally, from every data object, the Extractor sub-module extracts the needed attributes (based on the event definition). Both data objects and their attributes will be used in the following steps for lattice construction.

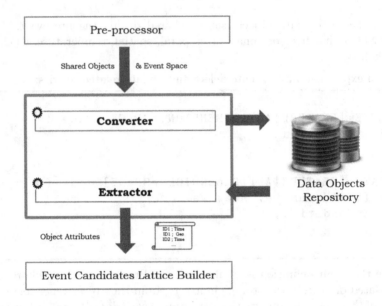

Fig. 4. The attribute extractor module

Lattice Construction: In this step, an event agent processes the previously extracted attributes, and data objects into lattice attributes and objects, in order to generate one output: the lattice. The *Feature-centric Event Lattice Builder* is the FCA backbone. It integrates the four step process of FCA clustering described in Sect. 3.2. To do so, we define lattice attribute types in Definition 4. These types will be used when defining the lattice attributes (cf. Definition 5). Lattice attributes, as defined here, ensure that any given object/attribute can be represented in the FCA formal context. Therefore, any object having attributes can be properly integrated in the clustered data set. This allows the event detection process of eVM to be generic and applicable in various contexts (e.g., social events, sensor events). Finally, for object/lattice attribute mapping, we define a binary cross rule denoted BXR (cf. Definition 6). This process is repeated for each event detection run.

Definition 4. *lat is a lattice attribute type representing an interval* $[a, b[$ *where* $lat : \langle a, b, T \rangle$, *where:*

- *a is the lower boundary value*
- *b is the upper boundary value*
- *T is a value representing the period having a primitive data type of either integer or float, such that:*
  - $dt(a) = dt(b) \in ADT$ *and*
  - $b = a + T$. ∎

Definition 5. *A lattice attribute, denoted* **la**, *is defined as a 4-tuple* $la : \langle f, eSpace, lat, y \rangle$ *where:*

- $f \in F$ is the event feature mapped to lattice attribute la
- $eSpace$ is the event space in which the detection will take place
- lat (cf. Definition 4) is the lattice attribute type
- y is a granularity | $y \in f.G$ and

$$lat.T = \begin{cases} y & \text{if } f.interval = True \\ 0 & Otherwise \end{cases}$$

$lat.a = so_i.v_j$, where:

- $so_i \in eSpace.SO$ and
- $(v_j \in so_i.V) \wedge (dt(v_j).f = f)$. ∎

For example, from the fire event example, we can find the following lattice attributes:

- Time intervals
- Geo locations
- Temperature intervals
- CO2 intervals
- Smoke existence (or not)

Definition 6. *A binary cross rule, denoted as **BXR**, is defined as a function that maps a shared object x to its respective lattice attribute y where $x.v_i \in x.V$:*

$$BXR = \begin{cases} 1 & \text{if } (y.lat.T = 0 \wedge y.lat.a = x.v_i) \vee \\ & (y.lat.T \neq 0 \wedge x.v_i \in [y.lat.a, y.lat.b[) \\ 0 & Otherwise \end{cases}$$

∎

Then the *Feature-centric Event Lattice Builder* constructs the FED (Feature-centric Event Detection) formal context, denoted ***ffc*** (cf. Definition 7). Once the ***ffc*** is created, formal concepts (cf. Fig. 6) are extracted and a lattice (cf. Fig. 7) is generated. This process is described in steps 2–4 of Sect. 3.2. This lattice is called an Event Candidate Lattice, where each node is a potential feature-centric event. Figure 5 illustrates the inner composition of the Event Candidates Lattice Builder module.

Definition 7. *A FED Formal Context, denoted **ffc**, is defined as a 6-tuple ffc :* $\langle eSpace, F, f_{LAG}, X, Y, I \rangle$, where

- $eSpace$ is the event space in which the detection takes place
- F is the set of one event features
- f_{LAG} is the function that generates the lattice attributes, described in Algorithm 1
- $X = eSpace.SO$ is the set of shared objects

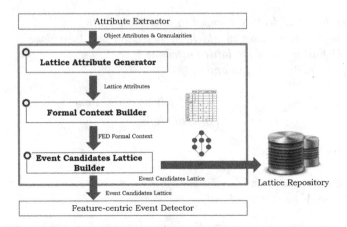

Fig. 5. The event candidates lattice builder module

Table 3. Fire event formal context example

	Time		Geo		Temp		CO2		Smoke	
	[9–9:15[	[9:15–9:30[	Loc1	Loc2	[20–40[	[40–60[	[250–350[	[350–450[	Yes	No
O1	X		X			X		X	X	
O2	X		X			X		X	X	
O3	X		X			X		X	X	
O4		X		X	X		X			X
O5		X		X	X		X			X
O6		X		X		X	X			X

- $Y = \bigcup_{i=0}^{|X.V|-1}\{la_i\}$ *is the set of lattice attributes* | $X.V = \bigcup_{\forall so \in X}\{so.V\}$ *is the union of all attribute values from the shared objects in eSpace*
- I *is a BXR(x,y) where* $x \in X \wedge y \in Y$. ∎

To follow up with the fire event example, Table 3 illustrates how lattice attributes (columns) are mapped to incoming sensor observations (rows) using the binary cross rule in the FED formal context.

The example in Table 3 shows six sensor observations (objects), mapped to their respective attributes. For instance, observation 1 has a timestamp value between 9 and 9:15 AM, therefore it is mapped to the lattice attribute [9–9:15[. This observation is taken from a sensor deployed in Loc1 and has a temperature reading that is included in the [40–60[degrees Celsius interval. Moreover, the other five observations are also mapped to their corresponding attributes using the binary cross rule. This represents the (FED) formal context in this scenario.

In Algorithm 1, we detail the lattice attribute generation process. This starts by extracting all object attribute values (lines 5–11). If the value is mapped to a feature that is generated as an interval (e.g., *time*), the algorithm calls the Create-Intervals function (lines 19–23). If not (e.g., *social*), the algorithm generates a lattice attribute type having a null period and creates the corresponding lattice attribute (lines 13–18). This step allows the creation of generic lattice attributes from various features, thus providing extensibility (criterion 2). Algorithm 2 details the Create-Intervals function. This process extracts all values related to the same feature (lines 4–9), orders them (line 10), selects a minimum and a maximum value (lines 11–12), and creates periodic intervals starting from the minimum to the maximum value (lines 14–22). The period is calculated based on the chosen feature granularity (line 15). This makes the detection more user-centric (criterion 1). Finally, the result is added the output of Algorithm 1.

Algorithm 1. Lattice Attribute Generation (cf. Definition 7 - f_{LAG})

```
 1  Input: eSpace
 2  Output: RES                                 // List of all lattice attributes
 3  VAL = new List()                            // Shared Objects attribute values list
 4  PD = new List()                             // Processed event features list
 5  foreach so ∈ eSpace.SO do
 6      foreach v ∈ so.V              // This loop extracts all object attribute values
 7      do                           from all objects in eSpace and stores them in
 8          if (v ∉ VAL) then        the VAL list
 9              VAL←v
10      end
11  end
12  foreach v ∈ VAL do
13      if (not dt(v).f.Interval)           // If the value is not generated as an
14      then                                   interval
15          lat ← LAT(v, lat.a + lat.T, 0)
16          la ← LA(dt(v).f, eSpace, lat, dt(v).f.g)         // Create la with lat.T=0
17
18          RES ← la
19      else
20          if (dt(v).f ∉ PD) then                          // Call
21              RES ← (Create-Intervals(VAL,v,PD, eSpace))    Create-Intervals
22                                                            function
23      end
24  end
25  return RES
```

Algorithm 2. Create-Intervals

---

1 Input: VAL, v, PD, *eSpace*   // Input provided by Algorithm 1, line 21
2 Output: LAI   // Generated lattice attributes intervals
3 int i = 0
4 TEMP = new List()   // Temporary object attribute list
5 **foreach** $val \in VAL$ **do**
6   **if** *(dt(val).f == dt(v).f)*   // Extract all object attribute
7   **then**   values having the same feature
8 | TEMP ← *val*   as v and store them in TEMP
9 **end**
10 $Order_{ascending}(TEMP)$ // Order TEMP ascending
11 min ← TEMP.get(0)   // min is the first element of TEMP
12 max ← TEMP.get($|TEMP| - 1$)   // max is the last element of TEMP
13 lat ← LAT()
14 **while** *(lat.b < max)* **do**
15   lat ← LAT(min, lat.a + (i+1) × lat.T, dt(v).f.g)
16   **if** *(lat.b > max)*   // This loop
17   **then**   creates
18 | lat.b ← max   intervals of
19   la ← LA(dt(v).f, *eSpace*, lat, dt(v).f.g)   period lat.T =
20   LAI ← la   f.g (feature
21   i++   granularity)
22 **end**
23 PD ← dt(v).f   // Add feature to the list of processed features
24 return LAI

Formal Concepts of "Fire Event Example"
1	<{O1,O2,O3}, {[9:00 - 9:15[,Loc1,[40 - 60[,[350 - 450[,Yes}>
2	<{O4,O5}, {[9:15 - 9:30[,Loc2,[20 - 40[,[250 - 350[,No}>
3	<{O6}, {[9:15 - 9:30[,Loc2,[40 - 60[,[250 - 350[,No}>
4	<{O1,O2,O3,O6}, {[40 - 60[}>
5	<{O4,O5,O6}, {[9:15 - 9:30[,Loc2,[250 - 350[,No}>
6	<{O1,O2,O3,O4,O5,O6}, {}>
7	<{}, {[9:00 - 9:15[,[9:15 - 9:30[,Loc1,Loc2,[20 - 40[,[40 - 60[,[250 - 350[,[350 - 450[,Yes,No}>

Fig. 6. Fire event formal concepts

Figure 6, shows the formal concepts extracted from the fire event FED formal context using the FCA operators. And finally, the generated lattice is shown in Fig. 7 where nodes are potential fire events.

Event Detection: The *Feature-centric Event Detector* module uses the previously generated lattice, an event detection rule (an operator that checks the lattice nodes in order to find the targeted events), and the features in order to detect feature-centric events (cf. Definition 8). We define a default event detec-

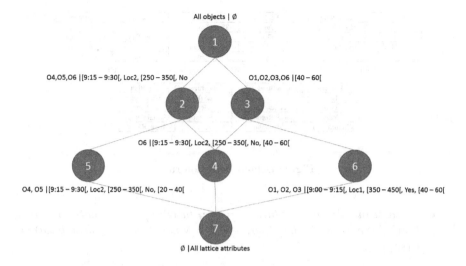

Fig. 7. Fire event lattice

tion rule, as a set of lattice attributes that comply with the two conditions mentioned in Definition 8. The rule is extensible, thus allowing the integration of multiple event features (e.g., *Time, Geo-location, Social, Topic*), each represented by the corresponding lattice attribute. This rule (cf. Fig. 8(a)) uses the selected key features in order to target the related feature-centric events. For example, the rules illustrated in Fig. 8(b), (c), (d), and (e) detect user, geo, topic, and time-centric events respectively (in a social event detection context). For instance, Fig. 8b detects social events (anytime, anywhere, and any topic) of each specific person. Fig. 8c is another example of a detection rule where the key feature is the geo location, i.e., this rule detects events that happened anytime, with any person, and concerning any topic at a specific location. These rules reflect the targeted events'definitions and are extensible. Features can be added/removed (e.g., temperature, CO_2 levels, smoke) for specific event detection needs (e.g., fire events could be defined as events having high temperature, CO2 and smoke at any time and place. Therefore, node 6 in Fig. 7 is detected as a fire event). Finally, for testing purposes, developers can change/add detection rules using the *Rule Selector* module. Since the lattice is not affected by the rule change, only the event detection step is repeated based on the new detection rule.

Definition 8. *A feature-centric Event, denoted **fce**, is a Formal Concept defined as a 4-tuple* $fce : \langle ffc, central_F, A, B \rangle$, *where:*

- *ffc is a FED Formal Context (Definition 7)*
- *$central_F$ is the set of selected key features* $|central_F \subseteq ffc.F$
- *A is a set of data objects* $| A \subseteq ffc.X$
- *B is a set of lattice attributes* $| B \subseteq ffc.Y$ *where* $\forall b_i, b_j \in B \wedge i \neq j$:
  - **Condition 1:** $b_i.f \neq b_j.f$

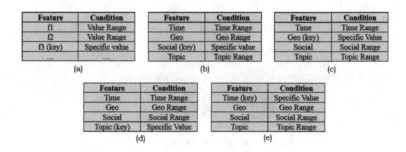

(a)   (b)   (c)

(d)   (e)

Fig. 8. Default detection rule

- **Condition 2:** if $b_i.f.label = c_f.label | \forall c_f \in central_F$, then $dist(b_i.lat.a, so_j.v_k) = 0 \mid \forall so_j \in A \wedge \forall v_k \in so_j.V$, $dt(b_i.lat.a) = dt(so_j.v_k)$.

∎

Finally, Fig. 9 details the interaction between the Rule Selector module and the Feature-centric Event Detector module. A detection rule change can be requested through the Event Query Language. One can create a new rule, select, or update an existing one. Based on the developer's choice, the Rule Validator sub-module checks the syntax of the newly created or updated rule prior to storage. If one decides to select an existing rule, the Rule Selector sub-module returns the chosen rule. In both cases, the event detection is repeated using the chosen rule and new results are generated.

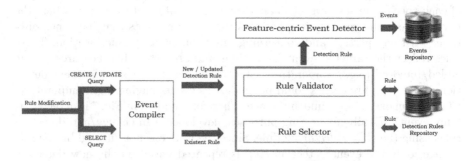

Fig. 9. The rule selector module

The syntax to create, select, or update a detection rule (DR) is described below:

```
CREATE DR <dr_id> (
[f_id =] {<feature_id>},
[key_id =] {<feature_id>},
);

SELECT DR <dr_id> FROM Detection_Rules_Repository(
WHERE Condition,
);

UPDATE DR <dr_id> (
SET [f_id =] {<feature_id>},
WHERE Condition,
);
```

Where f_id is a list of features (identified by their respective ids) that are considered by the detection rule, and key_id is the set of key features considered by the detection rule.

Finally, event spaces (eSPACE) do not initially contain any data objects (SO) because they are created prior to the event detection process. Once the latter occurs, an insert query adds each data object to its corresponding event space. The following describes the syntax of an insert query:

```
INSERT INTO [eSPACE] <id> (
{<value>},
WHERE Condition,
);
```

4 Implementation and Evaluation

We wanted to evaluate how generic and re-usable the framework is. Therefore, we instantiated from the eVM two applications: the first, for social event detection, and the second for conflict event detection. These two contexts provide a variety of features each (social features, conflict related features), and different data objects (multimedia shared data in the social context, news stories in the conflict events context). More particularly, we aimed to validate the components related to the Event Detection part of the eVM framework (Event Detector modules cf. Fig. 3). In order to do so, we measured, for each application (social/-conflict event detection), the event detection accuracy (with the integration and adaptation of FCA). We also evaluated the algorithm's performance based on execution time and memory consumption. The objectives of the experimentation were the following: (i) show that the approach is re-usable (criterion 1) by detecting feature-centric events in different contexts (e.g., social event detection, conflict event detection); (ii) measure the impact of domain specific expertise (criterion 2) on the detection accuracy (regarding the choice of event features and granularities in the event definition phase) while reducing human intervention (criterion 9); (iii) demonstrate that eVM is multi-modal (criterion 6) and multi-source (criterion 7) by measuring the impact of adding various features on the performance from various data sources; and (iv) proving that eVM is accurate when given (by the domain experts) optimal features/ granularities, and event definition. We do not aim at comparing accuracy results with other works since the objective is to provide a re-usable, easy to integrate, accurate backbone for event detection, that can be adjusted/configured by domain experts (i.e., by defining events, detection rules, features, granularities). We do not present here the evaluation results related to (i) the incremental processing (even though FCA is incremental [39]); (ii) evolution; (iii) extensibility; and (iv) ease of integration. This will be presented in a separate work.

Regarding the implementation, we developed a platform that integrates a front end mobile application and a cloud-based back end. In order to have a mobile application for both Android and iOS users, we used Visual Studio and Xamarin[2], a mobile development platform that allows building native mobile applications from a shared C# code. Xamarin also provides the following features: (i) complete binding for the underlying SDKs for both Android and iOS; (ii) Objective-C, Java, C, and C++ interoperability; (iii) mobile cross platform support; (iv) advanced base class library; and (v) Xamarin.Forms maximizes code sharing for cross-platform development, Xamarin.iOS and Xamarin.Android provide direct access to platform-specific APIs (cf. Figs. 10 and 11).

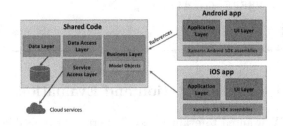

Fig. 10. Cross platform coding   **Fig. 11.** Xamarin forms architecture

We detail next the implementation of the two instances of the framework. Then, we show the evaluation of performance and accuracy by detecting feature-centric events in different contexts:(i) Social (e.g., wedding, birthday) Event Detection; and (ii) Conflict (e.g., wars, protests) Event Detection. In each context, we detail the corresponding dataset, and performance and accuracy tests/results.

---

[2] https://developer.xamarin.com.

4.1 Instantiating Applications from the eVM Framework

As previously mentioned, we created two instances from the eVM framework. The first is dedicated to feature-centric social event detection. In this case, the targeted events were defined based on various social-related features such as participants and topics in addition to time and location. The data objects, i.e., the event related data that need to be clustered into feature-centric events, were considered to be photos and videos taken/shared during the social events by participants. The second instance is designed for feature-centric conflict event detection. In this case, the targeted events were defined based on time, geo-location, and a set of contextual features such as the actors (i.e., aggressor, defender), the news source that covered the conflict, the conflict type (e.g., protest, war, planned attack), and finally the number of casualties. The data objects are considered news stories regarding the targeted conflict events.

4.2 Algorithm Evaluation

To evaluate the re-usability of our approach, we detected feature-centric events in different contexts. First, we detail our experimentations and results for the social event detection application (Test 1), then for the conflict event detection application (Test 2). The performance tests were conducted on a machine equipped with an Intel i7 2.60 GHZ processor and 16 GB of RAM. The aim was to test the performance of our eVM Event Detector algorithm.

Test 1: Application to Social Event Detection ReSEED Dataset: To evaluate the detection results, we used the ReSEED Dataset, generated during the Social Event Detection' of MediaEval 2013 [34]. It contains real photos crawled from Flickr, that were captured during real social events which are heterogeneous in size (cf. Fig. 12) and in topics (e.g., birthdays, weddings). The dataset contains 437370 photos assigned to 21169 events. In our evaluation, we used three event features: *time, location, and social,* since ReSEED photos have *time, geo, and social* attributes. In ReSEED, 98.3% of photos contain *capture time* information, while only 45.9% of the photos have a *location*. We had to select photos having these attributes from the dataset. This left us with 60434 photos from the entire dataset. In ReSEED, the ground truth used for result verification assigns photos to social events. Since, our approach is focused on feature-centric events (in this experimentation, user-centric events), we modified the ground truth to split the social events into their corresponding user-centric events. Since the splitting is based on the event features, we need to specify the feature granularities during the process. The latter are not specified in ReSEED, therefore we chose the lowest granularity values: *day* for *time, street* for *geo,* and *photo creator name* for *social.* The ground truth refactoring process is described in Fig. 13. First, we extracted the photos of each event in the ground truth. Second, we used the *timestamps of photo capture* to group photos by *day.* Third, we split the resulting clusters into distinct groups based on *street* values. Finally, the result was further split based on distinct *photo creators.*

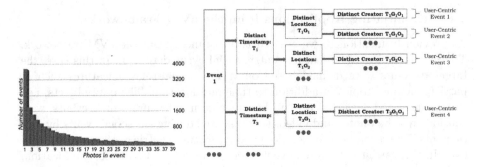

Fig. 12. ReSEED photo distribution

Fig. 13. Refactoring ground truth

Performance Evaluation: We considered two criteria for this task: (i) total execution time and (ii) memory overhead.

Use Cases: The performance is highly affected by the number of photos, generated attributes, and clusters. We noticed that granularities *day* for *time* and *street* for *geo* generate more clusters and attributes than any other granularity combination. Therefore, we used *day* and *street* to test the prototype's performance in three worst case scenarios:

- Case 1: We selected the biggest event (1400 photos) as input. We varied the number of photos progressively from 1 to 1400. Since all photos are related to one event, the number of detected clusters should be one.
- Case 2: We extracted 400 events each having exactly one photo. We varied the number of photos from 100, 200, 300 to 400. The number of generated clusters for each iteration should be 100, 200, 300, and 400 respectively.
- Case 3: The goal is to test with as many photos as possible related to different events. We varied the number of photos from 15000, 30000, 45000 to 60434. Since thousands of events contain only one or two photos per event (worst case scenario), this case will generate the most clusters.

Results and Discussion: In Cases 1 and 2 (Fig. 14a and b), where the number of photos does not exceed 1400 and 400 respectively, the total execution time is quasi-linear. However, in Case 3 (Fig. 14c), we clustered the entire dataset (60434 photos). The total execution time tends to be exponential, in accordance with the time complexity of FCA. When considering RAM usage, we noticed a linear evolution for the three cases (Fig. 14d, e, and f). RAM consumption is significantly higher in Case 2, where we generated 400 clusters, than in Case 1, where we generated one cluster. In Case 3, RAM consumption is the highest because both the number of photos at the input, and the number of generated clusters (detected events) were the highest. Other tests were conducted, Fig. 15 (left) shows that low granularities (e.g., day) consume more execution time than high ones (e.g., year). This is due to the generation of more lattice attributes

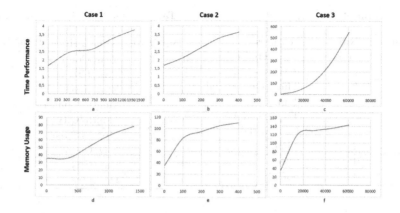

Fig. 14. Performance results

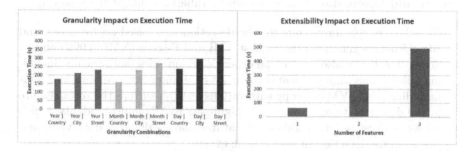

Fig. 15. Granularity and extensibility impact

and clusters. In addition, Fig. 15 (right), shows that considering more features in the processing is also more time consuming. Nonetheless, the evolution from one to three features remains quasi-linear, making the process extensible.

Accuracy Evaluation: We chose to consider the criteria proposed by MediaEval for clustering quality evaluation. We calculated the F-score, based on the Recall (R) and Precision (PR), and the Normalized Mutual Information (NMI) using ReSEED's evaluation tool. These criteria are commonly adopted in information retrieval and social event detection. A high F-score indicates a high quality of photo to user-centric event assignment while NMI will be used to measure the information overlap between our clustering result and the ground truth data. Therefore, a high NMI indicates accurate clustering result.

Use Cases: Since we considered the *time*, *geo*, and *social* features, we identified all possible combinations of the detection rule (see Table 4). In order to test granularity impacts, Table 5 sums up the different granularity combinations. When applying detection rules to granularity combinations, we get 63 use cases. We measured for each one the NMI and F-Score.

Table 4. Detection rule

Combination	Number of features	Features considered in the detection rule
1	3	Time, Geo, Social
2	2	Time, Geo
3		Time, Social
4		Geo, Social
5	1	Time
6	1	Geo
7	1	Social

Table 5. Granularity combinations

Combination	Granularities: Time/Geo
1	Year/Country
2	Year/City
3	Year/Street
4	Month/Country
5	Month/City
6	Month/Street
7	Day/Country
8	Day/City
9	Day/Street

Results and Discussion: Results shown in Table 6, highlight the following:

(i) *Detection rule/features impact:* The detection rule based on *time*, *geo*, and *social* features generates the highest NMI and F-score (NMI: 0.9999 and F-Score: 0.9995). It also exceeds all other detection rules (e.g., the one including solely *time* and *geo* features) in every granularity combination. This underlines that eVM can cope with various features such as the *social* feature in the detection task. Moreover, it highlights eVM's multi-modality, which allows the integration of additional features (having different datatypes) and the accurate detection of user-centric events.

(ii) *Granularity impact:* The results improve, when the clustering is based on granularities closer to the ones used in the ground truth. For example, in the case of granularities *year, country*, the F-Score achieved based on *time* and *geo* features is 0.1911, but for the detection rule that considers only the *social* feature the F-Score is higher: 0.5376. This is because the granularities for *time* and *geo* are the most general (*year and country*). Therefore, the impact factor of granularities is more important than that of the number of features considered in the detection rule. Some rules can exceed others for specific granularity combinations (e.g., *Time Geo* exceeds *Time Social* and *Geo Social* for granularities Year/Month/Day-Street while Time Social exceeds the other two rules for Year/Month/Day-Country). The best result can be achieved by considering the maximal number of features having correct granularities. This indicates that the granularities should not be fixed for all scenarios. When given the best granularities, our approach detects the user-centric events very accurately.

Test 2: Application to Conflict Event Detection ACLED Dataset: The aforementioned experiments targeted feature-centric social events which have features such as time, location, social (e.g., participants), and topics (e.g., birthday, marriage). In the following experiments, we targeted feature-centric conflict

Table 6. Clustering results

Detection rule	Measure	Granularities								
		Year			Month			Day		
		Country	City	Street	Country	City	Street	Country	City	Street
Time Geo Social	F-Score	0.6399	0.8180	0.8662	0.7964	0.8619	0.8948	0.9535	0.9742	**0.9995**
	NMI	0.9181	0.9602	0.9729	0.9549	0.9703	0.9789	0.9880	0.9938	**0.9999**
Time Geo	F-Score	0.1911	0.7678	0.8473	0.4943	0.8367	0.8821	0.8854	0.9542	0.9892
	NMI	0.7113	0.9475	0.9684	0.8707	0.9637	0.9759	0.9729	0.9894	0.9977
Time Social	F-Score	0.6245	0.6245	0.6245	0.7939	0.7939	0.7939	0.9534	0.9534	0.9534
	NMI	0.9143	0.9143	0.9143	0.9544	0.9544	0.9544	0.9879	0.9879	0.9879
Geo Social	F-Score	0.5085	0.7718	0.8357	0.5085	0.7718	0.8357	0.5085	0.7718	0.8357
	NMI	0.8742	0.9470	0.9653	0.8742	0.9470	0.9653	0.8742	0.9470	0.9653
Time	F-Score	0.0220	0.0220	0.0220	0.1399	0.1399	0.1399	0.7278	0.7278	0.7278
	NMI	0.3971	0.3971	0.3971	0.7069	0.7069	0.7069	0.9392	0.9392	0.9392
Geo	F-Score	0.0559	0.6958	0.8343	0.0559	0.6958	0.8343	0.0559	0.6958	0.8343
	NMI	0.5084	0.9241	0.9646	0.5084	0.9241	0.9646	0.5084	0.9241	0.9646
Social	F-Score	0.5376	0.5376	0.5376	0.5376	0.5376	0.5376	0.5376	0.5376	0.5376
	NMI	0.8755	0.8755	0.8755	0.8755	0.8755	0.8755	0.8755	0.8755	0.8755

events. The latter, have different features than social events (e.g., time, location, aggressor, defender, press news source, conflict type, casualties). This allowed to have a different event definition, as well as test the accuracy of the detection process in different contexts. For this purpose, we used the ACLED[3] (Armed Conflict Location & Event Data Project) Dataset. It is a disaggregated conflict collection, analysis and crisis mapping project. ACLED collects the dates, actors, types of violence, locations, and fatalities of all reported political violence and protest events across Africa, South Asia, South East Asia and the Middle East. Political violence and protest include events that occur within civil wars and periods of instability, public protest and regime breakdown. The dataset contains event records that span over years. We selected 49000 events, from the African subset, that date from April 1998 till January 2018. For each event, we generated shared objects having twelve attributes each (a object owner (press news source), a latitude, a longitude, a country, a region, a city, a street, a datetime, a conflict type, two actors, and the number of casualties). In total, we tested the accuracy of the detection process, the impact of the number of included features on the performance based on an input of 50000 shared objects related to the events mentioned above.

Performance Evaluation: To give a detailed view of the algorithm's performance, we measured the impact of: (i) the number of objects at the input; and (ii) the number of included features on the execution time.

Use Cases: The performance is affected by the input size, i.e., the number of objects to be processed, and the number of event features included in the clustering. Therefore, we experimented the following cases:

---

[3] https://www.acleddata.com.

– Case 1: We ran the detection module five times and measured the execution time of each run. Every iteration has the following configuration: (i) all seven features are considered in the clustering; and (ii) the granularity choices for time and geo-location are day and street respectively. The only variable is the input size. The first run processes 10000 objects, the second 20000, the third run 30000, then we considered 40000 in the fourth run, and finally 50000 in the last run.

– Case 2: We ran the detection module seven times and measured the execution time of each run. Every iteration has the following configuration: (i) the input size is the same, 50000 objects (the entire dataset); (ii) the granularity choices for time and geo-location are day and street respectively. In the first run, we only consider one feature (time) in the clustering. Then, for each iteration, we include one additional feature (e.g., time and geo-location in the second run). The last run includes all seven features (time, geo-location, press news source, actor 1, actor 2, conflict type, and casualties).

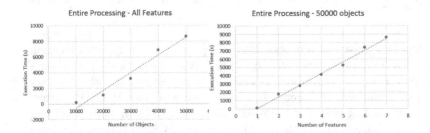

Fig. 16. Case 1 results   Fig. 17. Case 2 results

Results and Discussion: Figure 16 shows the impact of augmenting the size of the input on the algorithm's execution time (Case 1). We notice that the evolution of execution time is quasi-linear. Figure 17 shows the impact of including more features in the processing on the execution time (Case 2). The evolution is also quasi-linear in a worst case scenario from a granularity and input size point of view. We also analyzed the execution time of each step of the event detector (i.e., attribute extraction, lattice construction, event detection cf. Sect. 3). The highest cost in terms of execution time is related to the event detection step (not FCA computation), which consists of scrolling through the nodes of the lattice in order to select nodes that comply with the chosen feature-centric event definition. Results can be optimized by looking into better ways of scrolling through the lattice (graph analysis techniques). This will be conducted in a future work.

Accuracy Evaluation: We used the same metrics (F-Score & NMI) for accuracy evaluation. We tested the same granularity combinations shown in Table 5 for time and geo-location features. We did not vary the granularities of the five other features (press news source, actor 1, actor 2, conflict type, and casualties)

Table 7. Detection rule combinations

Combination	Number of features	Considered features
1	3	Time, Geo, Press news source
2	2	Time, Geo
3		Time, Press news source
4		Geo, Press news source
5	1	Time
6	1	Geo
7	1	Press news source

Table 8. ACLED accuracy results

Detection rule	Measure	Granularities											
		Year				Month				Day			
		Country	Region	City	Street	Country	Region	City	Street	Country	Region	City	Street
1	F-Score	0.3194	0.6181	0.6737	0.732	0.5012	0.7598	0.7873	0.8195	0.8439	0.9454	0.9528	**0.9631**
	NMI	0.8234	0.9287	0.9392	0.9515	0.9011	0.9604	0.9645	0.9701	0.9764	0.9923	0.9933	**0.9949**
2	F-Score	0.0254	0.3366	0.4385	0.5465	0.1789	0.6257	0.6771	0.7374	0.7252	0.921	0.9325	0.9477
	NMI	0.6652	0.8668	0.8885	0.9144	0.8263	0.9382	0.9461	0.9567	0.9598	0.9889	0.9904	0.9928
3	F-Score	0.2564	0.2564	0.2564	0.2564	0.429	0.429	0.429	0.429	0.8269	0.8269	0.8269	0.8269
	NMI	0.7892	0.7892	0.7892	0.7892	0.886	0.886	0.886	0.886	0.9742	0.9742	0.9742	0.9742
4	F-Score	0.232	0.5028	0.5751	0.6522	0.232	0.5028	0.5751	0.6522	0.232	0.5028	0.5751	0.6522
	NMI	0.7509	0.897	0.9133	0.9327	0.7509	0.897	0.9133	0.9327	0.7509	0.897	0.9133	0.9327
5	F-Score	0.001	0.001	0.001	0.001	0.0116	0.0116	0.0116	0.0116	0.291	0.291	0.291	0.291
	NMI	0.4261	0.4261	0.4261	0.4261	0.6671	0.6671	0.6671	0.6671	0.8941	0.8941	0.8941	0.8941
6	F-Score	0.0015	0.0865	0.1809	0.311	0.0015	0.0865	0.1809	0.311	0.0015	0.0865	0.1809	0.311
	NMI	0.4209	0.7434	0.7858	0.836	0.4209	0.7434	0.7858	0.836	0.4209	0.7434	0.7858	0.836
7	F-Score	0.1938	0.1938	0.1938	0.1938	0.1938	0.1938	0.1938	0.1938	0.1938	0.1938	0.1938	0.1938
	NMI	0.6913	0.6913	0.6913	0.6913	0.6913	0.6913	0.6913	0.6913	0.6913	0.6913	0.6913	0.6913

found in the ACLED dataset. As for the detection rule, we limited the combinations to three features: (i) time; (ii) geo-location; and (iii) press news source (chosen as central feature in this experimentation). This could be extended to include the actors of events (e.g., to detect the group most involved in armed conflicts), the number of casualties (e.g., to detect the deadliest conflict events), and the conflict types (e.g., to compare occurrence of protests between countries). The detection rule combinations are detailed in Table 7. Finally, we used the same tool (provided by the ReSEED work) for F-Score and NMI calculation. Accuracy results are detailed in Table 8.

Results and Discussion: Table 8 highlights the following:

(i) *Detection rule/features impact:* The detection rule 1 (based on *time, geo-location,* and *press news source* features) generates the highest NMI and F-score (NMI: 0.9949 and F-Score: 0.9631). It also exceeds all other detection rules (e.g., the one including solely *time* and *geo-location* features) in every granularity combination. This underlines that eVM can provide the appropriate way to conduct the clustering and integrate various datatypes related to a multitude of features due to its multi-modality. However, accuracy can still be improved, few objects were assigned to the wrong clusters. These errors can be minimized by including more features in the detection rule (i.e., the actors, conflict type, and casualties).

(ii) *Granularity impact:* We notice here (as we also did for the ReSEED dataset) that when the clustering is closer in terms of granularities to the ground truth, accuracy improves. For example, in the case of granularities *year, country*, the F-Score achieved based on *time* and *geo-location* features is 0.0254, but for the detection rule that considers only the *press news source* feature the F-Score is higher: 0.1938. This is because the granularities for *time* and *geo* are the most general (*year and country*). Some rules can exceed others for specific granularity combinations. The best result can be achieved by considering the maximal number of features having correct granularities. Finally, these results prove that eVM is re-usable since event detection accuracy was high in different event detection contexts.

5   Clustering Techniques

5.1   Clustering Techniques

Many works in different areas (e.g., information retrieval, event detection, image searching and annotation), have evolved around clustering techniques since their introduction in 1975 when John Henry Holland wrote Adaptation in Natural and Artificial Systems, a ground breaking book on genetic algorithms [18]. Unsupervised clustering is considered since in most cases, we detect events from raw data without prior knowledge on the occurring events. Clustering techniques are commonly grouped into four categories [38]:

Prototype-Based Clustering. A cluster is a set of objects that are closest (most similar) to the prototype that defines the cluster than to the prototype of any other cluster. A prototype can be the centroid or the medoid depending on the nature of the data (continuous attributes or categorical attributes). For continuous data, a centroid represents the object with the average (mean) values of all objects (points) in the cluster. As for categorical attributes, since a centroid is not meaningful, the prototype is often a medoid, the most representative point of the cluster. For many types of data, the prototype can be regarded as the most central point. Therefore, prototype-based clustering is commonly referred to as center-based clustering. For example, K-means [19] is a prototype-based clustering technique that groups objects based on a specified similarity measure (e.g., Euclidean distance, Manhattan distance, cosine similarity, Jaccard measure) and creates a set of K clusters represented each by a centroid. K-medoids [21] is another example of this clustering category. Instead of calculating means, actual points from the data are picked as representatives (prototypes) of the clusters. Points are associated to the clusters where they are most similar to the prototype. An iterative swapping process between prototypes and non prototype points is done as long as the quality of the clustering is improved.

These methods have low complexities for both time and space. But the algorithms attempt to find a predefined number of clusters (K): the final number of clusters should be known prior to clustering. In addition, for K-means, in order to start the clustering, the user has to choose initial cluster centers (centroids). This is a key step, if these centroids are chosen randomly clustering results can be poor.

Density-Based Clustering. A cluster is represented as a dense region surrounded by a low density region. Objects in the low density zones are considered noise while others in high density regions belong to the group limited by the region. For example, DB-Scan [33] produces a partitional clustering based on density measures. This method studies the neighborhood of each point, and partitions data into dense regions separated by not-so-dense regions. To do so, density at a point p is estimated by counting the points within a circle of center p and radius ϵ. Therefore, a dense region is a circle of radius ϵ containing a minimal number of points.

On one hand, DB-Scan determines automatically the number of clusters, is relatively resistant to noise, and can handle clusters of arbitrary sizes and shapes. On the other hand, since clustering is affected by the specified radius, DB-Scan loses accuracy when the clusters have widely varying densities. Also, with high-dimensional data, defining the densities becomes more difficult and more expensive (in term of computation time and space). Finally, points in the low-density areas are considered noise which means that not all input data will be present in the clusters.

Graph-Based Clustering. Data is organized in graphs/hierarchies where nodes are objects and connections among objects are represented as links connecting the nodes. Therefore, a cluster is defined as a connected component, a group of objects that are connected to one another but have no connections to objects from outside the group. For example, Agglomerative Hierarchical clustering is a graph-based clustering method [5]. First, each point is considered as a singleton cluster. Then repeatedly, the closest two clusters (based on similarity/dissimilarity matrices) are merged until a single all-encompassing cluster remains. Hierarchical clustering can also be divisive, this method is symmetrical to the agglomerative technique. In the divisive algorithm, all points are initially assigned to a single cluster and then based on similarity/dissimilarity measures the splitting into different clusters begins, until each point is assigned to a distinct cluster.

The added value of this method is that clusters are nested in a dendrogram (hierarchical structure) which offers a first level of semantic reasoning by exploiting the hierarchy and the inter-cluster relations. In contrast, the method has a high complexity in both time and space. All cluster merges are final, for high dimensional data such as photos, this is considered as a limitation. Since high dimensional data is more complicated, error correction if data is wrongly assigned to a certain cluster is a major issue.

Conceptual Clustering (Shared-property). A cluster is a set of objects that share some properties. For successful clustering, an algorithm would require a very specific definition of a cluster. This means that prior to the clustering, the shared properties that identify a cluster should be defined in order to generate a concept describing a cluster. The process of generating such clusters is called conceptual clustering. Formal Concept Analysis (FCA) is a conceptual, hierarchical clustering method [7,30]. It analyses data based on object/attribute relationships, extracts concepts, and finally orders them in a lattice. The advantage of having a lattice of formally defined Concepts is that it assures a more advanced level of semantic reasoning. In addition, FCA automatically generates a brief description for each cluster. Nonetheless, time and space complexities could cause concerns in some worst case scenarios where every data object forms a formal concept. In this case, exponential complexities become major technical difficulties.

Discussion: Table 9 shows a comparative summary of clustering techniques with respect to our defined, clustering-related, criteria (cf. Sect. 2). Prototype-based methods require excessive human intervention and the number of clusters prior to the processing. This is a major limitation in an event detection scenario where the total number of events is unknown prior to detection. In addition, these approaches are not multi-modal, multi-source nor incremental. Density-based methods detect automatically the number of final clusters, thus reducing human intervention but they are not multi-modal nor multi-source. These methods do not consider different types of data at once. In addition, clustering high dimensional data is complicated when relying on density measures. Graph-based (Hierarchical) clustering offers better semantic reasoning compared to the first

Table 9. Clustering technique comparison

Criterion	Clustering technique			
	Prototype-based [19,21]	Density-based [33]	Graph-based [5]	Conceptual [7,30]
Multi-modality	No	No	No	Yes
Multi-source	No	No	No	Yes
Incremental processing	No	No	No	Yes
Level of human intervention	High	Moderate	Moderate	Low
Predefined cluster number[a]	Yes	No	No	No

[a]This criterion states if the final number of clusters is required prior to clustering.

two techniques. It enables a first level of semantic-based processing by exploiting the hierarchy and inter-cluster relations. In addition, Hierarchical Clustering is accurate but remains highly expensive computation wise. Finally, Conceptual Clustering presents two main advantages. Firstly, incremental algorithms exist and offer lower complexities for time and space. Secondly, these techniques (e.g., FCA) offer two levels of semantic reasoning: (i) handling formal concepts as nodes, and (ii) generating an ordered lattice of concepts (nodes). Finally, FCA is multi-modal, dynamic, and multi-source.

6   Conclusion and Future Work

Event Detection is an essential part of many applications/services from various application domains (e.g., social, medical, home/building management, energy management, navigation systems, industry and manufacturing). All these approaches are task-centric, and designed for a specific application domain/purpose. However, these works do not share a common re-usable backbone for event detection that can be instantiated/used in different contexts. In this paper, we propose a generic framework, the event virtual machine (eVM), for feature-centric event detection. Our approach allows to target events using an SQL-like query language (eQL), thus creating a specific instance for each use case using the same framework. The detection part is based on Formal Concept Analysis (FCA), an incremental and dynamic clustering technique. We developed a prototype for testing purposes. The results show that our approach achieved high accuracy in most cases, especially when additional features (in addition to time and location) are considered. Results also proved that eVM is re-usable and multi-modal. As future work, we are investigating the detection of optimal granularities. We would also like to help improve the accuracy by automatically considering spatio-temporal distances between clusters and noise handling techniques. Finally, we want to extend the event language even more in order to test the extensibility, ease of integration, and evolution of the eVM framework based on the detection needs.

Acknowledgments. We thank Dr. Gilbert Tekli and Dr. Yudith Cardinale for their valuable feedback and input. We would also like to thank Anthony Nassar for his remarkable contribution in developing the mobile application used for the experimentation of this work.

References

1. Agarwal, Y., Balaji, B., Gupta, R., Lyles, J., Wei, M., Weng, T.: Occupancy-driven energy management for smart building automation. In: Proceedings of the 2nd ACM Workshop on Embedded Sensing Systems for Energy-Efficiency in Building, pp. 1–6. ACM (2010)
2. Aggarwal, C.C.: Managing and Mining Sensor Data. Springer, New York (2013). https://doi.org/10.1007/978-1-4614-6309-2

3. Allan, J., Papka, R., Lavrenko, V.: On-line new event detection and tracking. In: Proceedings of the 21st Annual International ACM SIGIR Conference on Research and Development in Information Retrieval, pp. 37–45. ACM (1998)

4. Bahrepour, M., Meratnia, N., Havinga, P.J.: Sensor fusion-based event detection in wireless sensor networks. In: 6th Annual International Mobile and Ubiquitous Systems: Networking & Services, MobiQuitous 2009, pp. 1–8. IEEE (2009)

5. Berkhin, P.: A survey of clustering data mining techniques. In: Kogan, J., Nicholas, C., Teboulle, M. (eds.) Grouping Multidimensional Data, pp. 25–71. Springer, Heidelberg (2006). https://doi.org/10.1007/3-540-28349-8_2

6. Buckman, A., Mayfield, M., BM Beck, S.: What is a smart building? Smart Sustain. Built Environ. **3**(2), 92–109 (2014)

7. Burmeister, P.: Formal concept analysis with ConImp: introduction to the basic features. Technische Universität Darmstadt, Fachbereich Mathematik (2003)

8. Cao, L., et al.: Image annotation within the context of personal photo collections using hierarchical event and scene models. IEEE Trans. Multimedia **11**(2), 208–219 (2009)

9. Chen, L., Roy, A.: Event detection from flickr data through wavelet-based spatial analysis. In: Conference on Information and Knowledge Management, pp. 523–532 (2009)

10. Choi, V.: Faster algorithms for constructing a concept (galois) lattice. In: Clustering Challenges in Biological Networks, p. 169 (2006)

11. Chong, C.Y., Kumar, S.P.: Sensor networks: evolution, opportunities, and challenges. Proc. IEEE **91**(8), 1247–1256 (2003)

12. Compton, M., et al.: The SSN ontology of the W3C semantic sensor network incubator group. Web Semant.: Sci. Serv. Agents World Wide Web **17**, 25–32 (2012)

13. Cooper, M., et al.: Temporal event clustering for digital photo collections. ACM Trans. Multimedia Comput. Commun. Appl. **1**(3), 269–288 (2005)

14. Cui, J., et al.: EasyAlbum: an interactive photo annotation system based on face clustering and re-ranking. In: Conference on Human Factors in Computing Systems, pp. 367–376. ACM (2007)

15. Doolin, D.M., Sitar, N.: Wireless sensors for wildfire monitoring. In: Smart Structures and Materials 2005: Sensors and Smart Structures Technologies for Civil, Mechanical, and Aerospace Systems, vol. 5765, pp. 477–485. International Society for Optics and Photonics (2005)

16. Ganter, B., Wille, R.: Formal Concept Analysis: Mathematical Foundations. Springer, Heidelberg (2012)

17. Hanlon, M., Anderson, R.: Real-time gait event detection using wearable sensors. Gait & Posture **30**(4), 523–527 (2009)

18. Holland, J.H.: Adaptation in natural and artificial systems: an introductory analysis with applications to biology, control, and artificial intelligence. University of Michigan Press (1975)

19. Jain, A.K.: Data clustering: 50 years beyond k-means. Pattern Recogn. Lett. **31**(8), 651–666 (2010)

20. Jasiewicz, J.M., et al.: Gait event detection using linear accelerometers or angular velocity transducers in able-bodied and spinal-cord injured individuals. Gait & Posture **24**(4), 502–509 (2006)

21. Koperski, K., Adhikary, J., Han, J.: Spatial data mining: progress and challenges survey paper. In: ACM SIGMOD Workshop on Research Issues on Data Mining and Knowledge Discovery, pp. 1–10 (1996)

22. Labeodan, T., De Bakker, C., Rosemann, A., Zeiler, W.: On the application of wireless sensors and actuators network in existing buildings for occupancy detection and occupancy-driven lighting control. Energy Buildings **127**, 75–83 (2016)
23. Lee, J., Bagheri, B., Kao, H.A.: A cyber-physical systems architecture for industry 4.0-based manufacturing systems. Manuf. Lett. **3**, 18–23 (2015)
24. Leonardi, C., Cappellotto, A., Caraviello, M., Lepri, B., Antonelli, F.: SecondNose: an air quality mobile crowdsensing system. In: Proceedings of the 8th Nordic Conference on Human-Computer Interaction: Fun, Fast, Foundational, pp. 1051–1054. ACM (2014)
25. Li, S., Son, S.H., Stankovic, J.A.: Event detection services using data service middleware in distributed sensor networks. In: Zhao, F., Guibas, L. (eds.) IPSN 2003. LNCS, vol. 2634, pp. 502–517. Springer, Heidelberg (2003). https://doi.org/10.1007/3-540-36978-3_34
26. Mei, T., et al.: Probabilistic multimodality fusion for event based home photo clustering. In: International Conference on Multimedia and Expo, pp. 1757–1760 (2006)
27. Oeldorf-Hirsch, A., Sundar, S.S.: Social and technological motivations for online photo sharing. J. Broadcast. Electron. Media **60**(4), 624–642 (2016)
28. Papadopoulos, S., et al.: Cluster-based landmark and event detection for tagged photo collections. IEEE MultiMedia **18**(1), 52–63 (2011)
29. Park, S.C., Park, M.K., Kang, M.G.: Super-resolution image reconstruction: a technical overview. IEEE Sig. Process. Mag. **20**(3), 21–36 (2003)
30. Priss, U.: Formal concept analysis in information science. Arist **40**(1), 521–543 (2006)
31. Quack, T., Leibe, B., Van Gool, L.: World-scale mining of objects and events from community photo collections. In: International Conference on Content-Based Image and Video Retrieval, pp. 47–56 (2008)
32. Raad, E.J., Chbeir, R.: Foto2Events: from photos to event discovery and linking in online social networks. In: International Conference on Big Data and Cloud Computing, pp. 508–515. IEEE (2014)
33. Rehman, S.U., et al.: DBSCAN: past, present and future. In: International Conference on Applications of Digital Information and Web Technologies, pp. 232–238 (2014)
34. Reuter, T., et al.: Reseed: social event detection dataset. In: Conference on Multimedia Systems, pp. 35–40. ACM (2014)
35. Sakaki, T., Okazaki, M., Matsuo, Y.: Earthquake shakes Twitter users: real-time event detection by social sensors. In: Proceedings of International Conference on WWW, pp. 851–860. ACM (2010)
36. Sayyadi, H., et al.: Event detection and tracking in social streams. In: ICWSM (2009)
37. Sheba, S., Ramadoss, B., Balasundaram, S.R.: Event detection refinement using external tags for flickr collections. In: Mohapatra, D.P., Patnaik, S. (eds.) Intelligent Computing, Networking, and Informatics. AISC, vol. 243, pp. 369–375. Springer, New Delhi (2014). https://doi.org/10.1007/978-81-322-1665-0_35
38. Tan, P.N., et al.: Introduction to Data Mining. Pearson Education India (2006)
39. van der Merwe, D., Obiedkov, S., Kourie, D.: AddIntent: a new incremental algorithm for constructing concept lattices. In: Eklund, P. (ed.) ICFCA 2004. LNCS (LNAI), vol. 2961, pp. 372–385. Springer, Heidelberg (2004). https://doi.org/10.1007/978-3-540-24651-0_31

40. Wahyudi, W.A., Syazilawati, M.: Intelligent voice-based door access control system using adaptive-network-based fuzzy inference systems (ANFIS) for building security. J. Comput. Sci. **3**(5), 274–280 (2007)

41. Welch, J., Guilak, F., Baker, S.D.: A wireless ECG smart sensor for broad application in life threatening event detection. In: 26th Annual International Conference of the IEEE Engineering in Medicine and Biology Society, IEMBS 2004, vol. 2, pp. 3447–3449. IEEE (2004)

42. Wille, R.: Restructuring lattice theory: an approach based on hierarchies of concepts. In: Rival, I. (ed.) Ordered Sets, pp. 445–470. Springer, Dordrecht (1982). https://doi.org/10.1007/978-94-009-7798-3_15

43. Wong, J.K., Li, H., Wang, S.: Intelligent building research: a review. Autom. Constr. **14**(1), 143–159 (2005)

44. Wu, Y.H., Miller, H.J., Hung, M.C.: A GIS-based decision support system for analysis of route choice in congested urban road networks. J. Geog. Syst. **3**(1), 3–24 (2001)

45. Yu, L., Wang, N., Meng, X.: Real-time forest fire detection with wireless sensor networks. In: Proceedings of 2005 International Conference on Wireless Communications, Networking and Mobile Computing, vol. 2, pp. 1214–1217. IEEE (2005)

46. Zampolli, S., Elmi, I., Ahmed, F., Passini, M., Cardinali, G., Nicoletti, S., Dori, L.: An electronic nose based on solid state sensor arrays for low-cost indoor air quality monitoring applications. Sens. Actuators B: Chem. **101**(1–2), 39–46 (2004)

Interactive Exploration of Subspace Clusters on Multicore Processors

The Hai Pham[1], Jesper Kristensen[2], Son T. Mai[2,3,5(✉)], Ira Assent[2],
Jon Jacobsen[2], Bay Vo[4], and Anh Le[5]

[1] Bach Khoa University (BKU), Ho Chi Minh City, Vietnam
hai.phamthepth@gmail.com
[2] Aarhus University, Aarhus, Denmark
{mtson,ira}@cs.au.dk
[3] University of Grenoble Alpes, Grenoble, France
mtson@univ-grenoble-alpes.fr
[4] Ho Chi Minh University of Technology (HUTECH), Ho Chi Minh City, Vietnam
vd.bay@hutech.edu.vn
[5] University of Transport, Ho Chi Minh City, Vietnam
anhlvq@gmail.com

Abstract. The PreDeCon clustering algorithm finds arbitrarily shaped
clusters in high-dimensional feature spaces, which remains an active
research topic with many potential applications. However, it suffers from
poor runtime performance, as well as a lack of user interaction. Our new
method AnyPDC introduces a novel approach to cope with these problems by casting PreDeCon into an anytime algorithm. In this anytime
scheme, it quickly produces an approximate result and iteratively refines
it toward the result of PreDeCon at the end. AnyPDC not only significantly speeds up PreDeCon clustering but also allows users to interact
with the algorithm during its execution. Moreover, by maintaining an
underlying cluster structure consisting of so-called *primitive clusters* and
by block processing of neighborhood queries, AnyPDC can be efficiently
executed in parallel on shared memory architectures such as multi-core
processors. Experiments on large real world datasets show that AnyPDC
achieves high quality approximate results early on, leading to orders of
magnitude speedup compared to PreDeCon. Moreover, while anytime
techniques are usually slower than batch ones, the algorithmic solution
in AnyPDC is actually faster than PreDeCon even if run to the end.
AnyPDC also scales well with the number of threads on multi-cores
CPUs.

Keywords: Subspace clustering · Anytime clustering
Interactive algorithm · Active clustering

1 Introduction

Clustering is an important data mining task that separates objects into groups
(clusters) such that objects inside a group are more similar to each other than

© Springer-Verlag GmbH Germany, part of Springer Nature 2018
A. Hameurlain et al. (Eds.): TLDKS XXXIX, LNCS 11310, pp. 169–199, 2018.
https://doi.org/10.1007/978-3-662-58415-6_6

those in different groups. Traditional clustering algorithms find clusters in the full dimensionality (all attributes) of the data. However, when the dimensionality increases, distances between data objects are increasingly similar [13]. As a consequence, meaningful clusters no longer exist in full dimensional space, rendering these techniques inappropriate for high dimensional data.

Due to this *curse of dimensionality* problem [13,31], finding subspace clusters in subspace projections of the high dimensional space has become an active research topic with many application domains [31,47]. Most subspace clustering techniques use a fixed dissimilarity assessment, where a range query for each object determines its so-called neighborhood under the same distance function, e.g. Euclidean Distance in SUBCLU [33]. PreDeCon [6], however, takes a fundamentally different approach in that it locally evaluates the relevance of dimensions for each point. Specifically, each point is assigned a so-called *subspace preference vector* indicating the distribution of its local neighborhood along each dimension. These vectors are then used to weigh distance calculations to reflect the neighborhood structure. Finally, clusters are detected similarly as in the seminal DBSCAN approach [11], using this weighted similarity assessment.

A major drawback of PreDeCon is that it needs to calculate the neighborhoods of all objects both under the Euclidean distance and the weighted distance function, thus resulting in $O(n)$ queries to be made, where n is the total number of points. And it is a very expensive task, especially for large datasets.

In our conference publication [32], we propose an efficient algorithmic solution for the PreDeCon clustering model, called *anytime PreDeCon* (AnyPDC), with the following attractive characteristics:

- **Anytime.** AnyPDC is an anytime algorithm that allows for user interaction. It quickly produces an approximate result and then continuously refines it until it reaches the same result as PreDeCon. At any time while being executed, it can be suspended to examine intermediate results and resumed again to further improve results. This scheme is highly useful when coping with large datasets under arbitrary time constraints, and when exploring the data. To the best of our knowledge, AnyPDC is the first *anytime* extension of PreDeCon.
- **Efficiency.** Most indexing techniques for speeding up range query processing suffer from performance degeneration in high dimensional data due to the curse of dimensionality problem mentioned above. By using the Euclidean distance as a filter for the weighted distance, AnyPDC significantly reduces the runtime for calculating the neighborhoods of objects. Moreover, it iteratively learns the clustering structure to avoid neighborhood queries of objects that do not contribute to the determination of the final result. Consequently, it produces the same result as PreDeCon with substantially fewer range queries. As a result, AnyPDC is more efficient than PreDeCon.

In this journal version, we further extend the algorithm AnyPDC [32] into a parallel technique on multicore processors for further enhancing its performance while still retaining the anytime and efficiency properties described above.

- **Parallelization.** By maintaining a global graph structure representing local cluster information and processing queries in blocks, AnyPDC can be efficiently parallelized on shared memory architectures such as multicore processors. It scales very well with the number of threads, thus further enhancing the performance by exploiting the characteristics of modern hardware. To the best of our knowledge, AnyPDC is the first parallel extension of PreDeCon. Moreover, it is a *work-efficient* parallel method which runs faster than PreDeCon on a single thread due to its efficient query pruning scheme. In additional, AnyPDC still retains its anytime properties, thus making it both a parallel and anytime technique at the same time.

The rest of this paper is organized as follows. In Sect. 2, we briefly summarize the algorithm PreDeCon and review the background on anytime processing. The algorithm AnyPDC and its parallel version called AnyPDC-MC are described in Sect. 3. Experimental results are shown and discussed in Sect. 4. We briefly discuss related work in Sect. 5 before concluding our paper in Sect. 6.

2   Background

2.1   The Algorithm PreDeCon

In contrast to other subspace clustering techniques which cluster with a fixed distance measure for all points, the general idea of PreDeCon [6] is to determine a subspace preference vector for each point to weigh the impact of attributes on the distance assessment locally. Formally, we assume a dataset D with n points in m attributes $A = \{A_1, \cdots, A_m\}$. PreDeCon starts by assessing the neighborhood around each point using a range parameter $\epsilon \in \mathbb{R}$:

Definition 1 *(ϵ-neighborhood). The ϵ-neighborhood of a point p under Euclidean distance d is defined as $N_\epsilon(p) = \{q \in D \mid d(p,q) \le \epsilon\}$.*

In PreDeCon, each point p is assigned a so-called *subspace preference vector* which captures the main direction of the local neighborhood of p over the set of attributes with respect to a variance threshold $\delta \in \mathbb{R}$ and a predefined constant $\kappa \gg 1$.

Definition 2 *(Subspace preference vector). The subspace preference vector of a point p is defined as $\overline{w}_p = \{w_1, \ldots, w_d\}$ where*

$$w_i = \begin{cases} 1 \ if \ Var_{A_i}(N_\epsilon(p)) > \delta \\ \kappa \ otherwise \end{cases}$$

where $VAR_{A_i}(N_\epsilon(p))$ is the variance of $N_\epsilon(p)$ along attribute A_i:

$$Var_{A_i}(N_\epsilon(p)) = \frac{\sum_{q \in N_\epsilon(p)} (d(\pi_{A_i}(p), \pi_{A_i}(q)))^2}{|N_\epsilon(p)|}$$

where $\pi_{A_i}(p)$ is the projection of p onto A_i.

PreDeCon uses preference vectors as attribute weights when calculating the distance among points. If a dimension of a point p is *specific* (its variance is smaller than δ), this dimension has a higher weight.

Definition 3 *(Preference weighted distance). The preference weighted distance between two points p and q is defined as:*

$$d_{pref}(p, q) = max\{d_p(p, q), d_q(p, q)\}$$

where $d_p(p, q)$ is the preference weighted distance between p and q wrt. the preference weighted vector $\overline{w}_p$:

$$d_p(p, q) = \sqrt{\sum_1^d w_i \cdot d(\pi_{A_i}(p), \pi_{A_i}(q))^2}$$

Definition 4 *(Preference dimensionality). Preference dimensionality of a point p, denoted as $Pdim(p)$, is the number of attributes A_i with $Var_{A_i}(N_\epsilon(p)) \leq \delta$.*

$Pdim(p)$ reflects the number of attributes in which $N_\epsilon(p)$ is compact. Only points with low $Pdim$, using a threshold $\lambda \in \mathbb{N}$, are considered as members of clusters:

Definition 5 *(Preference weighted neighborhood). The preference weighted neighborhood of a point p is defined as $P_\epsilon(p) = \{q \in D \mid d_{pref}(p, q) \leq \epsilon\}$.*

Following the density-based clustering paradigm, points are assessed as core, border or noise depending on their (local) density with respect to threshold $\mu \in \mathbb{N}$.

Definition 6 *(Core, border and noise points). A point p is a core point, denoted $core(p)$, if $|P_\epsilon(p)| \geq \mu \ \wedge \ Pdim(p) \leq \lambda$. If p is not a core point but one of its preference weighted neighbors is one and $Pdim(p) \leq \lambda$ then it is a border point. Otherwise, it is a noise point.*

Clusters are built from core points using the notions of density-reachability and density-connectedness that reflect the intuition of points that reside within a contiguous dense area.

Definition 7 *(Density-reachability). Point p is directly density-reachable from q, denoted $p \triangleleft q$, iff $core(q) \ \wedge \ p \in P_\epsilon(q) \ \wedge Pdim(p) \leq \lambda$.*

Definition 8 *(Density-connectedness). Two points p and q are density-connected, denoted $p \bowtie q$, iff there is a sequence of points $(x_1, \ldots, x_m)$ where $\forall x_i : core(x_i)$ and $p \triangleleft x_1 \triangleleft \cdots \triangleright x_m \triangleright q$.*

A cluster is then defined as a set of density-connected points.

Definition 9 *(Cluster). A cluster is a maximal set C of density-connected points in D.*

Algorithmically, PreDeCon [6] constructs clusters by randomly selecting a point p and calculating $P_\epsilon(p)$. If p is a core, it is assigned a cluster label and all of its neighbors are used as seeds for expanding the current cluster until no more point is found. Then, PreDeCon continues by choosing another unprocessed point until all points are processed.

In terms of complexity, this algorithm requires $O(n)$ range queries under Euclidean distance for calculating the preference weighted vectors of all points and $O(n)$ preference weighted neighborhood queries for calculating the neighborhoods of points. Moreover, propagating cluster labels from points to their neighbors requires $O(n^2)$ time in the worst case. In total, this results in $O(dn^2)$ time complexity.

2.2 Anytime Algorithms

The above algorithm is, like most clustering algorithms, a *batch* algorithm, i.e., it only provides a single result at the end. This poses an issue given the high runtime of the clustering algorithm, especially when facing additional resource constraints, applications that require exploration of different parameter settings or data samples or other user interaction requirements. In particular, interaction requires the ability to examine intermediate clustering results to determine whether to continue the run for further refined clustering results.

Anytime algorithms [65] return approximate results whose quality is improved over time. They have proven useful for many time consuming applications, e.g., robotics [66], computer vision [35,40,44], artificial intelligence [29], or data mining [5,36–39,42,43,45,51,53,58].

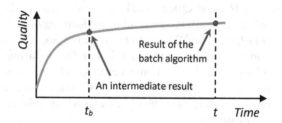

Fig. 1. A progress of an anytime algorithm

Figure 1 illustrates the idea with processing in an anytime fashion. Generally, an anytime algorithm can be interrupted at arbitrary points in time to review approximate results, e.g., at time t_b. The longer it is run, the better the quality of the result is expected to be. At time t, the result quality corresponds to that of the batch algorithm. After t, the quality of result may increase more. [65], describe some important properties of anytime algorithms, including that their final results should be similar to those of the batch ones and that final cumulative runtimes of anytime techniques are usually higher than for batch ones, ideally with as little extra overhead as possible.

3   The Algorithm AnyPDC

In this section, we present our algorithm anytime PreDeCon, AnyPDC. First, we introduce some important base concepts in Sect. 3.1, followed by a description of the proposed algorithm AnyPDC itself in Sect. 3.2, and of a parallel version, called AnyPDC-MC targeting multicore processors in Sect. 3.3. The algorithm is analyzed in Sect. 3.4.

3.1   General Ideas

Range Query Processing. We start from the observation that PreDeCon performance depends heavily on the neighborhood queries for constructing clusters; thus, enhancing them is crucial for speeding up PreDeCon. To this end, we establish the following relationship between Euclidean distance and the preference weighted distance.

Lemma 1. *For all points p and q, $d(p, q) \leq d_{pref}(p, q)$.*

Proof. Straightforwardly from Definition 3 using $\forall w_i : w_i \geq 1$(Definition 2).

Lemma 1 allows us to use Euclidean distance (ED), supported by any indexing technique for high dimensional data (or simply a sequential scan), as a filter for range queries. $P_\epsilon(p)$ can be directly obtained from $N_\epsilon(p)$ by calculating the preference weighted distance $d_{pref}(p, q)$ for all $q \in N_\epsilon(p)$. If $d_{pref}(p, q) \leq \epsilon$, q will belong to $P_\epsilon(p)$. This *filter-and-refinement* scheme may substantially reduce neighborhood computation time, thus speeding up the algorithm as a whole.

Block Processing of Range Queries. Block processing is a key idea to support parallel processing of AnyPDC. Instead of performing only one range query at a time as in PreDeCon, AnyPDC selects α points and performs the range query for several points, storing results in a buffer for updating the clustering structure in parallel as well. This scheme helps avoid synchronization among threads, thus enhancing the scalability of AnyPDC wrt. the number of threads. Block processing also reduces the overhead of the anytime scheme by decreasing the maximal number of iterations to n/α.

Summarization. PreDeCon expands a cluster by examining each point and its neighbors sequentially. This scheme incurs many redundant distance calculations, which decreases the performance of algorithm. Moreover, it is highly sequential and does not work well in parallel. AnyPDC, in contrast, follows a completely different approach inspired by [41]. It starts by summarizing points into homogeneous groups which we call *local clusters*. Each consists of a core point p as its representative and points that are density-connected. If they are all directly density-reachable from p, we call the local cluster a primitive cluster:

Definition 10 *(Local cluster and primitive cluster). A set S of points is called a local cluster of a point p, denoted $lc(p)$ iff p is a core and $\forall q \in S : p \bowtie q$. We also call $lc(p)$ a primitive cluster iff p is a core and $\forall q \in S : p \lhd q$.*

Lemma 2. *All points p_i with $p_i \in lc(q)$ for some q belong to the same cluster.*

Proof. Directly follows from Definitions 7 to 9.

Following Lemma 2, we only need to connect local clusters to produce the final clustering result of PreDeCon. Since the number of local clusters is much smaller than the number of points, label propagation time is greatly reduced. This leads to a reduction of the number of synchronizations required for propagating labels as compared to the original PreDeCon algorithm. Moreover, it allows AnyPDC to iteratively update the cluster structure in the block query processing scheme described above since we only need to update the connection between local clusters after additional queries. Summarization also provides a way to reduce the number of queries needed to determine a cluster as we shall see in the next sections.

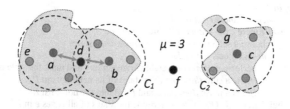

Fig. 2. General ideas of AnyPDC

Figure 2 illustrates these main ideas. Assume that we have three local clusters a, b, and c after the three first neighborhood queries. To determine cluster labels, we only need to label a, b, and c themselves. The remaining points can be labeled using the labels of their representatives. Next, we select d and g, query their neighborhoods, and find that d is a core. Following Lemma 3, a and b belong to the same cluster. From Lemma 4, we have that b and c cannot be connected. Thus, we detect the two clusters C_1 and C_2 exactly as PreDeCon does but using only 5 queries instead of 15 queries in total.

Lemma 3. *If $a \in lc(p) \cap lc(q)$ and $core(a)$, all points in $lc(p) \cup lc(q)$ belong to the same cluster.*

Proof. For any points $x \in lc(p)$ and $y \in lc(q)$, we have $p \bowtie x$ and $y \bowtie q$. Since a is a core object and $a \in lc(p) \cap lc(q)$, we have $a \bowtie x$ and $a \bowtie y$. Thus, x and y belong to the same cluster (Definitions 8 and 9).

Lemma 4. *Given two primitive clusters and their representatives p and q, if $d(p,q) > 3\epsilon$, p and q cannot be directly density-reachable under the d_{pref} distance.*

Proof. Let u and v be arbitrary points inside $lc(p)$ and $lc(q)$, respectively. We have $d(p,q) \leq d(p,u) + d(q,u)$ and $d(q,u) \leq d(q,v) + d(u,v)$ (triangle inequality).

Therefore, $d(u,v) \geq d(p,q) - d(p,u) - d(q,v)$. Since $d(q,u) \leq \epsilon$ and $d(q,v) \leq \epsilon$, we have $d(u,v) > \epsilon$. Due to Lemma 1, $d_{pref}(u,v) > \epsilon$, i.e., u and v are not directly density-reachable following Definition 7.

In the next section, we introduce the AnyPDC algorithm (non-parallel version). In Sect. 3.3, we discuss the parallelization of AnyPDC on shared memory architectures.

3.2   The Algorithm AnyPDC

```
1    function C = AnyPDC (O, μ, ε, α, δ, λ)
2    input:     dataset O, and parameters μ, ε of DBSCAN
3               the query block size α, and parameters δ and λ of PreDeCon
4    output:    the final clustering result C
5    begin
6        /* step 1: summarization */
7        while there exist untouched objects in O do
8            choose a set S of α objects with untouched states
9            for all objects o in S do
10               perform preference weighted range query and mark the state of o
11               if o is a core then mark the states of its neighbors in Pₑ(o)
12               if o is a noise then put o and Pₑ(o) into the noise list L
13       build graph G=(V, E) and determine the states of all edges e in E
14       /* repeat until terminated */
15       while (true) do
16           /* step 2:  check stopping condition */
17           find all connected components of the yes edges of G
18           merge each connected component of G into a single local cluster
19           set the new state for each edge of the new graph G
20           b = check if the termination condition is reached
21           if b = true then break
22           /* step 3: select objects for range queries */
23           for all nodes v in V do
25               calculate the node statistic and the node degree for v
26           compute scores for all unprocessed objects in O using deg(v)
26           choose a set S of α objects with highest scores
27           /* step 4: updating graph and clusters */
28           for all objects o in S do
29               perform the preference weighted range query on the object o
30               update the states of o and its preference weighted neighbors Pₑ(o)
31               merge Pₑ(o) to all nodes that contain o if o is a core
32           change the states of all edges e in E
33           /* step 5: find border and noise points */
34           for all objects o in L do
35               check if o is a border or a noise point
36       return the final clustering result C
```

Fig. 3. Pseudocode for AnyPDC

As described above, the general idea of AnyPDC is reducing the number of range queries to improve efficiency. AnyPDC therefore maintains a state that describes the current knowledge about its clustering status as follows.

Definition 11 *(Point state). The state of a point p represents its current clustering status under the execution of AnyPDC and is denoted as state(p). If a query has been performed on p, state(p) is* processed. *Otherwise, it is* unprocessed. *In addition, the state of p can be* core, border, *or* noise *following Definition 6. The* untouched *state means that p is unprocessed and it is not inside the neighborhoods of any processed points.*

Using these states that compactly capture our knowledge about points with respect to the clustering structure, we describe our efficient algorithmic solution that we structure in five main steps (see also Fig. 3 for the entire pseudocode). Step 1 summarizes points into local clusters and creates a graph to represent the current knowledge of their clustering relationships. Step 2 checks the graph to see if the relationships between all clusters are known. If that is not the case, then Step 3 selects points for neighborhood queries that are expected to reveal most about the clustering structure. Step 4 performs these queries and updates the graph accordingly. Finally, Step 5 refines noise points.

Step 1: Summarization. In the beginning, all points are marked as *untouched.* For each point q, we additionally store its currently known number of neighbors, denoted as $nei(q)$. More precisely, if a range query is executed on a point p and $q \in P_\epsilon(p)$ then we increase $nei(q)$ by 1.

AnyPDC iteratively and randomly chooses α points with *untouched* state, performs neighborhood queries, and checks if they are core objects. If a point p is a core, it is marked as *processed-core* and all of its directly density-reachable points will be assigned to a local cluster (primitive cluster) with p as a representative. For all points $q \in P_\epsilon(p)$, if *state(q)* is *untouched*, it will be marked as *unprocessed-border* or *unprocessed-core* depending on the number of neighbors $nei(q)$ it currently has. If *state(q)* is *processed-noise*, it is marked as *processed-border* since it is surely a border of a cluster. If p is a core point either in *processed* or *unprocessed* state, its state remains. If p is not a core, it is marked as *processed-noise* and is stored in the so-called noise list L for a post processing step. Note that, if $Pdim(p) > \lambda$, p is surely a noise point following Corollary 1. In this case, we do not need to calculate $P_\epsilon(p)$ and thus save runtime.

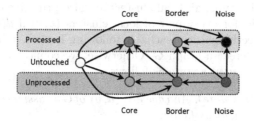

Fig. 4. Object transition state schema

The summarization process stops when all points are assigned to local clusters or to the noise list L, i.e., there are no *untouched* points left. Note that, in contrast to the original PreDeCon algorithm [6], *Pdim* calculation is performed only when it is required for calculating the preference weighted distances. We store all calculated preference weighted vectors for later re-use in preference weighted distance calculations in the interest of further runtime reduction.

Corollary 1. *If $Pdim(p) > \lambda$, p is a noise point.*

Proof. Directly inferred from Definition 6.

Figure 4 summarizes the transitions between states in AnyPDC. Please note that the state of a point p can only be changed from *noise* to *border* to *core* and from *unprocessed* to *processed* state.

Lemma 5. *For every point p, $state(p)$ only changes following the transition state scheme given in Fig. 4.*

Proof. All state transitions can be verified using Definitions 6 and 11. Assume that a point p is currently in *unprocessed-border* state. If a range query on p reveals that $P_\epsilon(p) < \mu$, then p is definitely a border point because one of its neighbors is a core point (since p was already labeled *border*). If $P_\epsilon(p) \geq \mu$, p is definitely a core point, making it *unprocessed-core* or *processed-core* depending on whether a range query on p has been executed. All other cases can be shown similarly.

As described in Sect. 3.1, we connect these local clusters to form the final clustering result. To do so, at the end of Step 1, AnyPDC builds a graph $G = (V, E)$ that represents the current connectivity state of all local clusters. G is used to analyze the potential cluster structures and to select points for neighborhood queries in the next steps.

Definition 12 *(Local cluster connection graph).* *The connection graph $G = (V, E)$ represents the current connectivity states of all local clusters, where each vertex $v \in V$ represents a local cluster and each edge $e = (u, v)$ of G represents the relationship between two vertices u and v.*

Note that, at this time, all local clusters are also primitive ones as stated in Definition 10. Thus, for every pair of local clusters (u, v) if $d(u, v) > \epsilon$, u and v will never be directly density-reachable following Lemma 4. Therefore, u and v will not be connected in graph G. Otherwise, we connect u and v in G and assign this edge $e = (u, v)$ a state as follows.

Definition 13 *(Edge state).* *An edge $e = (u, v)$ of G is assigned a state, denoted as $state(e)$ or $state(u, v)$ as follows:*

- *$state(u, v) = yes$ if $\exists a \in lc(u) \cap lc(v)$ with $core(a)$*
- *$state(u, v) = weak$ if $|lc(u) \cap lc(v)| \neq \emptyset$ and $state(u, v) \neq yes$*
- *$state(u, v) = unknown$ if $|lc(u) \cap lc(v)| = \emptyset$*

For simplicity, if $(u, v) \notin E$, we say that $state(u, v) = no$. The $state(u, v)$ therefore reflects whether points in $lc(u)$ and $lc(v)$ belong to the same cluster. If it is *weak*, u and v are more likely to be in the same cluster given our current knowledge than if it is *unknown*. As new knowledge on core points and local clusters is obtained, these states are updated accordingly. AnyPDC chooses points for neighborhood queries in the next steps based on these states.

Lemma 6. *For any pair of points p, q that belong to the same cluster in PreDe-Con, and their respective local clusters $lc(x), lc(y)$, there exists a path of edges in G that connects $lc(x)$ and $lc(y)$.*

Proof. Assume that there is no path in G that connects $lc(x)$ and $lc(y)$. Let $S = s_i$ be an arbitrary path of points that connects p and q. Let LS be a set of local clusters that contains all s_i. Due to Lemma 4, there must be two adjacent points s_i and s_{i+1} where $d(s_i, s_{i+1}) > \epsilon$ or $d_{pref}(s_i, s_{i+1}) > \epsilon$ (Lemma 1). This leads to a contradiction.

Due to Lemma 6, we only need to examine graph G to build clusters. At any point in time, clusters can be extracted by finding connected components of *yes* edges in the graph G following Lemma 7. All points in local clusters of a connected component are assigned the same cluster label. For brevity, we refer to them as connected components in the following.

Lemma 7. *Each connected component of yes edges of G is part of a subspace preference weighted cluster.*

Proof. Let $p \in lc(x)$ and $q \in lc(y)$ be two arbitrary points in the connected component C. Let $lc(x) = lc(c_1), \ldots, lc(c_t) = lc(y)$ be a path of local clusters that connects $lc(x)$ and $lc(y)$. Due to Definition 13, there is a set of core points that connects c_1 and c_t. Thus, p and q are density-connected. And they belong to the same cluster following Definition 9.

In the next step, AnyPDC iteratively selects points for neighborhood queries and updates the graph G until there are no *weak* or *unknown* edges left, using Lemma 8.

Lemma 8. *When all edges of G are in yes state, the overall cluster structure has converged.*

Proof. Due to Lemma 5, if a point p is a core either in *unprocessed* or *processed* states, it cannot change state to *border* or *noise*. Thus, for two arbitrary vertices u and v in G, if $state(u, v) = yes$, it would not change even if additional neighborhood queries were to be performed. Thus, the final clustering result has converged following Lemma 7.

Step 2: Checking Stopping Condition. At the beginning of step 2, AnyPDC finds all connected components of graph G. Then, it changes all *weak* and *unknown* edges in each connected component to *yes* state since their corresponding vertices belong to the same cluster as stated in Lemma 7.

If there are no more *weak* and *unknown* edges in G, the clustering result has converged as stated by Lemma 8. Thus, we can safely stop AnyPDC in the interest of saving runtime, even though many points may still be *unprocessed*. Besides its anytime property, this is a core difference between AnyPDC and the original algorithm PreDeCon which makes AnyPDC much more efficient than PreDeCon as mentioned already in in Sect. 3.1.

If there still exist *weak* or *unknown* edges, AnyPDC continues to perform range queries to determine the core properties of *unprocessed* points and to connect local clusters. Prior to this, however, AnyPDC merges all local clusters in each connected component into a single local cluster. One of the representatives of these local clusters will be selected as the representative for the whole group. The general purpose is to reduce the total number of graph nodes. Fewer vertices mean better performance and smaller overhead for maintaining the graph G, e.g., less time for finding connected components or merging clusters in the next steps. After merging local clusters, AnyPDC rebuilds the graph G following Lemma 9.

Lemma 9. *Given two connected components* $C = \{lc(c_1), \ldots, lc(c_a)\}$ *and* $D = \{lc(d_1), \ldots, lc(d_b)\}$ *of* G, *we have:*

- *Case A:* $\forall lc(c_i) \in C, lc(d_j) \in D : state(lc(c_i), lc(d_j)) = no \Rightarrow state(C, D) = no$
- *Case B:* $\exists lc(c_i) \in C, lc(d_j) \in D : state(lc(c_i), lc(d_j)) = weak \Rightarrow state(C, D) = weak$
- *Case C: otherwise* $state(C, D) = unknown$

Proof. Since there is no connection between $lc(c_i)$ and $lc(d_j)$, C and D will obviously have no connection. Thus, Case A holds. Moreover, according to Lemma 6, C and D surely belong to different clusters. Following Definition 13, $C \cap D \neq \emptyset$ since $\exists lc(c_i) \in C, lc(d_j) \in D : lc(c_i) \cap lc(d_j) \neq \emptyset$. Moreover, for all points p in $C \cap D$, p is not a core, since otherwise C and D will belong to the same connected component. Thus, case B holds. Case C is proven similarly.

In the next steps, AnyPDC selects points for neighborhood queries, determines the core properties of *unprocessed* points, and rebuilds graph G and the clusters until the termination condition of Lemma 8 is met.

Step 3: Select Objects. Selecting the right points for neighborhood queries is important if we want to reduce the number of queries. For example, if we choose points e and g in Fig. 2, the cluster structure does not change as a result of the respective queries. To avoid such waste, AnyPDC introduces an *active clustering* scheme [50] that iteratively analyzes the current result and scores all unprocessed points based on some statistical information, based on the connectivity states of G and based on their positions in G. Then, AnyPDC iteratively chooses the next α top scored points for neighborhood queries and updates to the clustering structure. Here we use the same heuristic proposed in [41] to select points.

Definition 14. *(Node statistic [41]). Given a node $u \in V$, the statistical information of u, denoted as $stat(u)$, is defined as follows:*

$$stat(u) = \frac{usize(u)}{|pclu(u)|} + \frac{|pclu(u)|}{n}$$

where $usize(u)$ is the number of points with unprocessed states in $pclu(u)$ and n is the total number of points.

Definition 15. *(Node degree [41]). Given a node u, the degree of u, denoted as $deg(u)$, is defined as follows:*

$$deg(u) = w(\sum\nolimits_{v \in N(u) \wedge state(u,v)=weak} stat(v))$$
$$+ \sum\nolimits_{v \in N(u) \wedge state(u,v)=unknown} stat(v) - \psi(u)$$

where $N(u)$ is the adjacent nodes of u in the graph G, $\psi(u) = 0$ if u is not a border node (i.e., u has some border points) otherwise $\psi(u)$ is the total number of weak and unknown edges of u.

The degree of a node u represents the uncertainty of the node u wrt. its local position. If u has high $deg(u)$, that means u is staying inside a highly uncertain area indicated by many undetermined edges around it. Thus, if we process u by casting a range query on it, u and its adjacent nodes v will more likely to be connected ($state(u,v) = yes$) or separated ($state(u,v) = no$). If $state(u,v) = weak$, u and v share some objects. Thus, they are more likely to be in the same clusters. In contrast, if $state(u,v) = unknown$, the connectivity status of u and v are less clear than the former case. To reflex these points, we assign to edges with *weak* states a higher weight $w = |V|$ and a lower weight to edges with *unknown* states (we use 1 here). Moreover, we want to postpone all border local clusters following Lemma 8 in order to bring AnyPDC to its final stage fasters.

Based on the node information of the graph G, we assign for each point p a score, denoted as $score(p)$, indicating how much it can affect the current cluster structure if we lauch a range query on p. The $score(p)$ is calculated based on the information of all local clusters that consist of p and the total number of neighbors of p we currently know. Note that, we want to choose objects lying inside high uncertainty areas, indicated by the sums of their node degrees. At the same time, we want to examine points with fewer neighbors since they may help to reveal more core points if range queries are performed on them.

Definition 16. *(Point score [41]). The score of an unprocessed point p, denoted as $score(p)$, is defined as follows:*

$$score(p) = \sum\nolimits_{u \in V \wedge p \in pclu(u)} deg(u) + \frac{1}{nei(p)}$$

where $nei(p)$ is the current number of neighbors of p.

Note that, we only choose *unprocessed* objects for querying. Otherwise, it is a waste since the query has been performed on all *processed* points.

Step 4: Update Cluster Structures. For any selected point p, we lauch a range query on it. If p is not a core object, we mark it as *processed-border* (because p is already a member of a cluster). Otherwise, p is a core and is marked as a *processed-core* point. All of its neighbors are reassigned new states following the transition state scheme in Fig. 4 as in Step 1. Then, we merge $P_\epsilon(p)$ into all local clusters that contain p.

Lemma 10. *If a point p and its neighbors are merged into local cluster $lc(q)$, they and $lc(q)$ belong to the same cluster.*

Proof. For arbitrary points $x \in P_\epsilon(p)$ and $y \in lc(q)$, we have $p \bowtie x$. Moreover, $q \bowtie p$ following Definition 10. Since p is a core object, we have $q \bowtie x$ due to Definition 8. Therefore, $P_\epsilon(p) \cup lc(q)$ is a local cluster and belongs to a cluster.

After merging, we must update the number of neighbors for all unprocessed objects q if $q \in P_\epsilon(p)$. Since $P_\epsilon(p)$ has been merged into a node v, the total number of unprocessed points inside v must be updated as well.

Since the states of points have been changed, the connections among vertices of G change as well. Thus, at the end of Step 3, G is updated locally, i.e., only edges related to changed clusters are examined. Their states are changed following Definition 13. Moreover, according to Lemma 11, if one of the two local clusters $lc(p)$ and $lc(q)$ has been fully processed, i.e., all their points are in *processed* states, and $state(lc(p), lc(q)) \neq yes$, then $lc(p)$ and $lc(q)$ belong to different clusters and their edge can be safely removed from G.

Lemma 11. *Given two local clusters $lc(p)$ and $lc(q)$, if the edge (p, q) is still in weak or unknown state but $lc(p)$ or $lc(q)$ are fully processed, i.e., queries are executed for all their points, $lc(p)$ and $lc(q)$ cannot be directly connected via their members.*

Proof. Assume that a node $lc(p)$ is fully processed and there exists a chain of density-connected core objects $X = \{p, \ldots, d, e, \ldots, q\}$ that connects the two core objects p and q (Definition 8) where $d \in lc(p)$ and $e \in lc(q)$. However, if e is a core object, we merged all of its neighbors including d into $lc(p)$. Thus, $state(lc(p), lc(q))$ must be *yes*. This leads to a contradiction.

Steps 2, 3, and 4 are repeated until the stopping criterion is met (Lemma 8). We now go on to the final Step 5.

Step 5: Determining Noise and Border Objects. The goal of this step is to examine the noise list L to see if points are true noise or border points. To do so, all points p in L are examined. If $Pdim(p) > \lambda$, p is surely noise according to Corollary 1. Otherwise, AnyPDC needs to check whether p is a border point of a cluster. If p is currently in *processed-border* state, it is already inside a cluster. Otherwise, AnyPDC checks all the preference weighted neighbors $q \in P_\epsilon(p)$. If

q is a core, then p will surely be a border point. If there is no core point in $P_\epsilon(p)$, additional range queries must be performed if one of p's neighbors is still in *unprocessed* state. If a core is found, p is a border point. Otherwise, p is surely noise.

In total, AnyPDC is an algorithm that is more efficient than PreDeCon, since it reduces the number of queries required to build clusters. Moreover, its *anytime* scheme provides a very useful way to cope with large datasets under arbitrary runtime constraints. In the following, we proceed to develop a parallel version of AnyPDC.

3.3 Parallelizing AnyPDC

We study how AnyPDC may exploit the inherent parallelism in virtually any modern hardware to further enhance performance. In this work, we focus on the multi-core processors which are ubiquitous in today's computers where multiple processors share are a common memory space. Assessing shared data is therefore very fast at internal memory speed instead of at slower network speeds, e.g., 19.2 GB/s for DDR4-2400 RAM vs. 1 GB/s LAN. However, maintaining data integrity and consistency as well as fully utilizing all processors are challenges that need to be addressed. Efficiently parallelizing AnyPDC is thus a non-trivial problem. For example, due to the overlap of the neighborhoods of points, two threads may concurrently assign different states for shared points at Step 1, thus leading to inconsistence. A naive approach is using locks to allow only one thread to change the states of points at a time. During that time, all other threads must remain idle waiting for the current one to finish before starting their turns. Another example is merging connected components. If we simply use one thread for processing one component, some threads might finish very fast and enter idle states doing nothing while others require much longer times. These problems thus significantly reduce the efficiency of the algorithm.

In this section, we introduce a scalable parallelization of AnyPDC on multi-core processors, called AnyPDC-MC. The general ideas are to reduce synchronizations among threads and to balance the workload of threads in each step of AnyPDC to fully utilize their processing power. The pseudocode of AnyPDC-MC is given in Fig. 5.

Step 1: Summarization. As briefly described in Sect. 3.1, we do not process each query using multiple threads since this requires many synchronizations. In AnyPDC-MC, each query is completely processed by a thread. Moreover, since the neighborhoods of points may overlap, a naive solution would simply use locks, but in the interest of efficiency we introduce a technique to mark the states of points instead such that locks are avoided.

First, all queries in a block S are processed by multiple threads. All the acquired results will be stored in a data buffer B for processing in the next phases. Moreover, we mark the states of all processed points p in S as *processed-core* if p is a core. If p has less than μ neighbors, it is assigned *processed-border* state if its current state is *unprocessed-border*. If p is in *untouched* state, it

```
1    function C = AnyPDC-MC ( O, μ, ε, α, δ, λ)
2    input:      dataset O, and parameters μ, ε of DBSCAN
3               the query block size α, and parameters δ and λ of PreDeCon
4    output:     the final clustering result C
5    begin
6       for all point o in O do                    ▶ parallel schedule dynamic
7          calculate the subspace preference vector of o
8       /* step 1: summarization */
9       while there exist untouched objects in O do
10         choose a set S of α untouched objects
11         for all objects o in S do               ▶ parallel schedule dynamic
12            perform preference weighted range query and mark the state of o
13         for all objects o in S do               ▶ parallel schedule dynamic
14            if o is a core then mark the states of its neighbors in P_ε(o)
15            for all p in P_ε(o) do change nei(p) using an atomic operation
16         for all objects o in S do
17            if o is a core then store lc(o) into the list of node V
18            if o is a noise then put o and P_ε(o) into the noise list L
19      for all node u in V do                      ▶ parallel schedule static
20         for all node v in V \ {u} do
21            if d(u,v) ≤ ε then add an edge (u, v) to E (d(u,v) ≤ d_pref(u,v))
22      for all edge e in E do                      ▶ parallel schedule dynamic
23         determine the states of e
24      /* repeat until terminated */
25      while (true) do
26         /* step 2: find subclusters and check for stopping condition */
27         find all connected components of the yes edges of G          ▶ *
28         for all pairs p of merged nodes do    ▶ parallel schedule dynamic
29            perform merging in parallel using boolean vectors
30         build the new graph G and calculate the states of its edges  ▶ *
31         b = check if the termination condition is reached            ▶ *
32         if b = true then break
33         /* step 3: select objects for range queries */
34         for all nodes v in V do                  ▶ parallel schedule static
35            calculate node statistic stat(v) for v
36         for all nodes v in V do                  ▶ parallel schedule static
37            calculate node degree deg(v) for v
38         for all unprocessed object o in O do  ▶ parallel schedule dynamic
39            calculate the object score score(o) for o
40         sort all unprocessed objects o according to score(o)         ▶ *
41         choose a set S of α objects with highest scores
42         /* step 4: update cluster structure */
43         for all objects o in S do               ▶ parallel schedule dynamic
44            perform preference weighted range queries on the object o
45         for all objects o in S do               ▶ parallel schedule dynamic
46            update the states of o and its neighbors P_ε(o)
47            for all p in P_ε(o) do change nei(p) using atomic operation
48         for pairs of merged object & node do ▶ parallel schedule dynamic
49            merge the preference neighborhood of object o into node u
50         for all changed edge e in E do          ▶ parallel schedule dynamic
51            update the state of e to reflex the new connectivity status
52         /* step 5: find border points or noise points */
53         for all objects o in L do               ▶ parallel schedule dynamic
54            check if o should be a border of a cluster
55      return the final clustering result C
```

Fig. 5. Pseudocode for AnyPDC-MC. (*) means that these parts may execute in parallel

will be changed to *processed-noise* state since it does not belong to any local clusters at this time. Obviously, there are no possible conflicts among threads here. However, since the neighborhood sizes are different, dynamic scheduling should be employed for workload balancing. At the end of this phase, a barrier allows threads to complete before moving on.

Second, for each point $p \in S$, a thread is used for assigning the state of all points $q \in N_\epsilon(p)$. In case p is a core, if q is *processed-noise*, it is changed to *processed-border* following the transition scheme in Fig. 4. Similarly, if q is in *untouched* state, it is switched to *unprocessed-border* state. Then, we increase the number of neighbors of q using an atomic operation to avoid inconsistencies when q is a neighbor of several points in S. After that, if $nei(q) \geq \mu$, $state(q)$ is changed to *unprocessed-core* if it is currently *unprocessed-border*. Similarly, a barrier is placed at the end for synchronizing, and dynamic scheduling is used for load balancing.

Third, for all points p in S, we do sequential work including: placing a point p in the noise list L or adding $clu(p)$ to the list of vertices of G, depending on query result. Then, AnyPDC repeats all these phases, processing other blocks until all objects are touched.

Instead of naive locking while assigning states to points, this scheme only requires an atomic operation for updating the numbers of neighbors of points, which is much faster than using locks, e.g., 200 times faster in OpenMP[1]. The sequential work is linear in the number of objects and is negligible compared to the overall runtime of the clustering algorithm. Thus, it does not affect the scalability much.

To determine the states of edges of G, naively constructing the graph G would be inefficient since the runtime varies greatly (cf. Definition 13), e.g., $O(1)$ for the *no* case and $O(n)$ for the other cases. Instead, we separate this process into two smaller phases: determining the edges and calculating their states.

First, we scan through the list of nodes V of G using multiple threads. For each $lc(p)$ and $lc(q)$, if $d(p,q) > 3\epsilon$, they are not connected. Otherwise, we add an edge between them to E. Its state is only be calculated in the next phase. A barrier is placed at the end for synchronizing.

Second, we determine the states of all edges E in parallel. Since, we exclude all *no* edges, the workload here is balanced better than in the naive way. We use dynamic scheduling also since the sizes of nodes are different.

Step 2: Checking Stopping Condition. To find connected components, if the number of nodes is small enough, e.g. less than 10000, a sequential algorithm is an efficient choice. Otherwise, any parallel technique for component finding can be employed, e.g., [12,34]. Since the number of nodes is typically small, even a sequential algorithm in this phase does not affect the scalability of AnyPDC-MC much.

Merging connected components is inefficient if we do it naively, e.g., using one thread for processing each component since the sizes of components are very

---

[1] http://openmp.org/wp/.

different. We therefore adopt a more efficient approach. First, we choose a node as a representative for each component. Second, we identify all pairs of nodes to be merged with this node inside each component. And third, we process all these pairs in all the components in parallel. However, since two nodes might be merged into the same node at the same time, we use a boolean vector to store the list of members in each node. Moreover, dynamic scheduling is also used since the sizes of nodes are different. Figure 6 illustrates the idea.

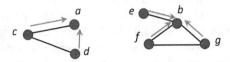

Fig. 6. Merging connected components. In the example, five pairs of nodes are merged as indicated by red arrows. (Color figure online)

Checking the stopping condition is simple. We only need to scan through the list of edges E in parallel if the number of nodes is relatively large, e.g., larger than 500. Otherwise, it should be done sequentially to limit overhead.

Step 3: Selecting Objects. Calculating the node statistics $stat(u)$ and node degrees $deg(u)$ can easily be done in parallel using static scheduling since the workload is balanced well. However, dynamic scheduling is required for calculating the point scores since each point may belong to many different nodes, thus making the workload imbalanced. For sorting the points, any parallel sorting technique can be used, e.g., [34].

Step 4: Updating the Clustering Structure. First, performing queries is similar to Step 1. Each query in a block S is processed by one thread and the result is stored in a buffer B for further processing. Then, B is used for assigning the states of objects instead of naively locking points.

Since each point p in S may belong to different primitive clusters, simply using one thread for merging $P_\epsilon(p)$ with all nodes containing p would obviously be inefficient due to imbalanced workloads. Thus, we collect all pairs of (point, node) that are to be merged together. Then multiple threads are used for building the results concurrently. Similar to Step 2, we use boolean vectors to store the memberships of points inside nodes. We illustrate the idea in Fig. 7. This approach balances the workload better, thus increasing the efficiency of AnyPDC-MC.

To update edge states in graph G after merging nodes, we extract the list of affected edges, i.e., edges with *weak* or *unknown* state first. Then the states of all selected edges are recalculated in parallel under a dynamic scheduling scheme since the sizes of nodes vary.

Step 5: Determining Noise and Border Points. We check all points in the noise list L to see if they are already registered as *unprocessed-border* or

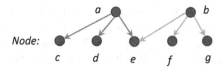

Fig. 7. Merging the neighborhoods of points a and b into primitive clusters c, d, e, f, and g. There are six merged pairs as indicated, e.g., (a, c) or (b, f).

processed-border with a linear scan over the noise list. Then, for all points p that are still undetermined ($state(p)$ is *processed-noise*), we process them in parallel checking whether they really belong to clusters. Since the amount of work may differ significantly between points, dynamic scheduling is employed in the interest of efficiency.

3.4 Algorithm Analysis

Lemma 12. *The final result of AnyPDC is identical to that of PreDeCon.*

Proof. AnyPDC produces clusters following the notion of PreDeCon as proven in Lemmas 2 to 11. Thus, when all queries are processed, it produces the same results as PreDeCon. Moreover, when all the edges of G are in the *yes* state, there cannot be further changes in graph G and thus the final clustering structure. Thus, stopping the algorithm at this point means that the result is the same as the final one, even though many points have not been subjected to a separate neighborhood query.

Please note that shared border points are assigned the label of one of its clusters based on the examining order of points in PreDeCon. The same is true for AnyPDC. Thus, shared border points might be assigned to different clusters in either algorithm (or in different runs of either algorithm with a different point order).

Lemma 12 describes an interesting property of AnyPDC in that it has the power of both approximation techniques (in terms of its anytime scheme) as well as of exact techniques if run intil convergence.

Analyzing time and space complexity of AnyPDC is a complicated task due to its anytime scheme. For simplicity, we only present here a worst case complexity.

Let n, v, and l are the number of points, number of initial graph nodes, and the size of the noise list L. Let b is the total number of iterations. The time complexity of AnyPDC is $O(v^2 n + b(v^2 + vn + n log n) + l\mu n)$ in the worst case. It consumes $O(v^2 + vn + l\mu)$ memory for storing the graph G and local clusters. Note that, v and b are much smaller than n and n/α for all real datasets in our experiments, respectively.

4 Experiments

All experiments are conducted on a Linux workstation with two 3.1 GHz Xeon CPUs and 64 GB RAM for each CPU using g++ 4.8.3 (-O3 flag) and OpenMP

3.1 [7]. We compare AnyPDC with the original PreDeCon algorithm on several real datasets including Electric Device (ED)[2], Brain-Computer Interfaces (BCI)[3], and Weather Temperature data (WT)[4]. They consist of 16637, 41680, and 566268 points in 96, 96 and 12 dimensions, respectively. Unless otherwise stated, we use default parameters $\kappa = 4$, $\delta = 0.0001$, $\lambda = 5$ (following the suggestions in [6]). Default value for μ is 5. The default block size is $\alpha = 128$.

4.1   Anytime PreDeCon

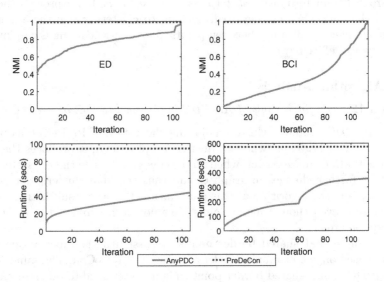

Fig. 8. NMI scores and runtimes of AnyPDC over the number of iterations (red line) in comparison with the final result of PreDeCon (indicated by dotted horizontal lines) for the datasets ED (left) and BCI (right) (Color figure online)

Anytime Performance. To assess the anytime property of AnyPDC, we use the results of PreDeCon as ground truth. At each iteration of AnyPDC, we measure how similar its intermediate result is compared to the result of PreDeCon using the Normalized Murtual Information (NMI) score [64]. NMI results are in $[0, 1]$ where 1 indicates perfect clustering results, i.e., AnyPDC produces the same results as PreDeCon.

Figure 8 shows the results of AnyPDC and PreDeCon for the ED ($\epsilon = 0.6$) and BCI ($\epsilon = 0.8$) datasets. As we can see, the NMI scores increase at each iteration and reach 1 at the end. That means the results of AnyPDC gradually come closer to those of PreDeCon and are identical at the end. For ED as an

---

[2] http://www.cs.ucr.edu/~eamonn/time_series_data/.

[3] http://www.bbci.de/competition/iii/.

[4] http://www.cru.uea.ac.uk/data/.

example, if we stop the algorithm at the first iteration, AnyPDC requires only 9.4 s compared to 93.3 s of PreDeCon, which is approx. 10 times faster. Even if it runs until the end, AnyPDC requires only 43.5 s (around 2.1 times faster than PreDeCon). This is interesting since the final runtimes of anytime algorithms are usually larger than those of the corresponding batch ones.

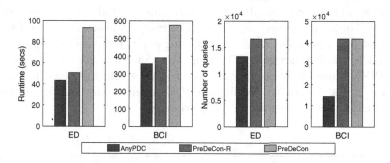

Fig. 9. Runtimes (left) and numbers of queries (right) of different algorithms for the dataset ED and BCI

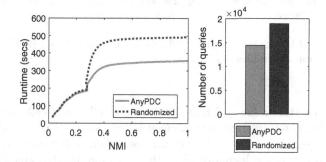

Fig. 10. Performance of AnyPDC's active clustering scheme for the dataset BCI ($\epsilon = 0.8$)

Why is AnyPDC Faster than PreDeCon? There are two reasons: (1) it has a more efficient neighborhood query scheme and (2) it uses fewer queries than PreDeCon due to its efficient query pruning scheme. To illustrate these two aspects, we create a modification of the original PreDeCon algorithm where we replace its query scheme with the one proposed in this work, named PreDeCon-R. Figure 9 shows the runtimes and numbers of queries of AnyPDC, PreDeCon, and PreDeCon-R for the two datasets ED and BCI with the same parameters as in Fig. 8. As we can see, due to its efficient querying scheme, PreDeCon-R is more efficient than PreDeCon even though they use the same numbers of queries, e.g., 50.7 s vs. 93.3 s for the dataset ED. With the same query scheme, AnyPDC is faster than PreDeCon-R since it uses fewer queries due to its query pruning scheme, e.g., AnyPDC uses 14345 queries compared to the 41680 queries of PreDeCon and PreDeCon-R.

Performance of AnyPDC's Active Clustering Scheme. Figure 10 (left) shows the NMI scores acquired at different time points during the execution of AnyPDC and a randomized technique. In this method, we randomly choose *unprocessed* objects for querying in Step 3 instead of the active selection scheme of AnyPDC. As we can see, randomly choosing points for querying requires much more time to reach a particular NMI score than for the *active clustering* scheme of AnyPDC. Moreover, more queries are needed to converge than in AnyPDC as shown in Fig. 10 (right). While AnyPDC requires only 14345 queries, the randomized strategy consumes 18947 queries.

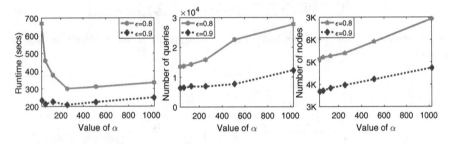

Fig. 11. Effect of block size α on the performance of AnyPDC on BCI dataset BCI, for different parameter values of ϵ

The Effect of Blocksize α. Figure 11 shows the effect of block size α (varied from 32 to 1024) on different performance aspects of AnyPDC on the dataset BCI for different values of ϵ. As we can see, when α increases, the number of initial graph nodes v increases since there is more overlap among nodes in Step 1 of AnyPDC. Similarly, large α means more points are selected at each iteration of AnyPDC. The more points we select in a block, the more redundant queries may occur since the clustering structure is not updated frequently. This increases the number of queries to produce clusters. However, the changes in both the number of queries and in number of nodes are small compared to the values of α. For example, with $\epsilon = 0.8$, when α is changed from 32 to 1024 (32 times), the number of nodes increases from 5139 to 6925 (1.34 times) and the number of queries grows from 13511 to 27648 (2.04 times).

Due to its anytime scheme, AnyPDC requires some overhead for scoring points and updating the clustering structure in each iteration. Thus, when α increases, the final cumulative runtime of AnyPDC decreases since it does not have to re-evaluate points frequently. Thus, the overhead is reduced. However, when α is large enough, redundant queries occur, making it a bit slower as we can see from Fig. 11. For $\epsilon = 0.8$ as an example, it takes AnyPDC 667.7 and 300.8 s when α grows from 32 to 256, respectively. However, when α is increased to 512 and 1024, the final runtime of AnyPDC slightly increases to 311.1 and 334.7 s, respectively.

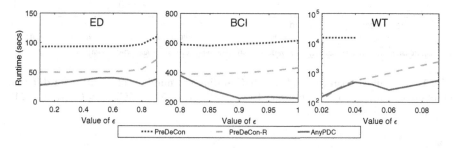

Fig. 12. Performance of different algorithms for the datasets ED, BCI, and WT, varying ϵ.

Performance Comparison. Figure 12 further compares the final cumulative runtimes of AnyPDC and of other techniques for different real datasets and values of ϵ. Note that we choose ϵ so that all points are noise in the beginning and increase it until most points are inside a single cluster. AnyPDC is much faster than the original PreDeCon algorithm, especially on large datasets. Taking the dataset WT as an example, with $\epsilon = 0.4$, PreDeCon takes 15019.1 s, while AnyPDC needs only 466.1 s, which is 32.2 times faster. Moreover, AnyPDC is also up to 4.31 times faster than PreDeCon-R on all datasets, even though it uses much fewer preference weighted neighborhood queries than PreDeCon-R. The reason is that both AnyPDC and PreDeCon must perform the neighborhood queries on Euclidean distance to calculate the preference weighted vectors for all points. Since this is expensive, it reduces the performance gap between AnyPDC and PreDeCon-R.

4.2   AnyPDC on Multicore Processors

AnyPDC on Multiple Threads. Figure 13 (left) shows the cumulative runtimes of AnyPDC during its execution with different numbers of threads. As we can see, the more threads are used, the faster the intermediate runtimes of AnyPDC-MC, which demonstrates the efficiency of our parallel strategy. Being an anytime algorithm and a parallel algorithm at the same time, AnyPDC-MC effectively uses multiple threads for acquiring (intermediate) results faster. When facing large datasets or time constraints, these properties are extremely useful. To the best of our knowledge, none of the existing PreDeCon variants has these properties. Moreover, AnyPDC-MC scales very well with the numbers of threads as additionally indicated in Fig. 13 (right). For the dataset ED as an

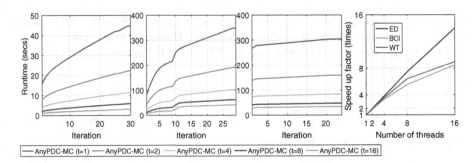

Fig. 13. Cumulative runtimes of AnyPDC at different numbers of iteration for different numbers of threads for the datasets ED ($\epsilon = 0.6, \alpha = 512$), BCI ($\epsilon = 0.8, \alpha = 1024$), and WT ($\epsilon = 0.06, \alpha = 1024$)

example, AnyPDC-MC acquires speedup factors of 1.96, 3.83, 7.46, and 14.02 using 2, 4, 6, 8, and 16 threads, respectively. For the datasets BCI and WT, the speedup factors are slightly worse since the neighborhood sizes of their points vary significantly, creating some imbalance of workloads.

We further study the performance of AnyPDC-MC on multiple threads in the next Sections. But first, we study the effect of block size α on the scalability of AnyPDC-MC.

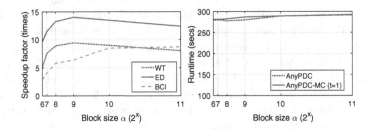

Fig. 14. The effect of block size α on the scalability of AnyPDC-MC

The Block Size α and the Scalability of AnyPDC-MC. Figure 14 (left) shows the scalability of AnyPDC-MC using 16 threads for the datasets ED ($\epsilon = 0.6$), BCI ($\epsilon = 0.8$), and WT ($\epsilon = 0.06$) when the block size α changes from 64 to 2048. Generally, it quickly increases with α and then slightly decreases when α is large enough. Taking the dataset ED as an example, the speedup factor is 9.54 when $\alpha = 64$, and reaches the highest value 14.02 when $\alpha = 512$. If we keep increasing α to 1024 and 2048, it slightly reduces to 13.48 and 12.40, respectively. The reason is that increasing α will increase the workloads in each iteration of AnyPDC-MC, e.g., more queries to be performed in Steps 1 and 3 and more nodes to be merged in Steps 2, 3, and 4. Using dynamic scheduling, each thread will have more work to do instead of idling, especially during Steps

2, 3 and 4 when local clusters are merged. This balances the workload better, thus leading to the improvement of the overall performance. However, when α is too large, there are more conflicts in Steps 1 and 3 since many points are selected for neighborhood queries at the same time. Moreover, balancing the workload becomes harder since the neighborhood sizes of points inside each block will vary more than when using smaller α. Accordingly, overall performance of AnyPDC-MC is slightly reduced in this case.

Figure 14 (right) shows the performance of the sequential algorithm AnyPDC and its parallel version AnyPDC-MC using a single thread only. Obviously, AnyPDC-MC is slightly slower than AnyPDC since it has some additional overhead for balancing the workload and for avoiding synchronizations, e.g., building list of merged pairs of local clusters in Steps 2 and 4. However, the difference is small (up to 2.5% of the final cumulative runtimes of AnyPDC only).

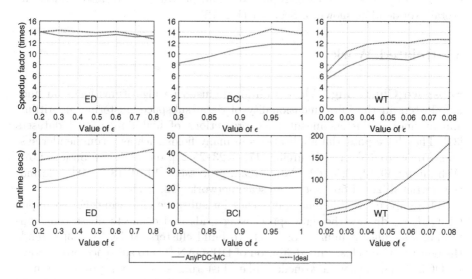

Fig. 15. The performance of AnyPDC-MC and an ideal algorithm for the datasets ED ($\alpha = 512$), BCI ($\alpha = 1024$), and WT ($\alpha = 1024$) using 16 threads.

Parallel Performance. Since AnyPDC-MC is the first parallel extension of PreDeCon, we further compare its performance to an ideal algorithm, denoted as *Ideal*. In this algorithm, we first use multiple threads for calculating the preference weighted vectors and dimensions for all points. Then, we calculate all the preference weighted neighboorhoods of all points using multiple threads. And last, we completely ignore the label propagation process of PreDeCon for labeling objects, which incurs many conflicts when done concurrently. Similar to AnyPDC-MC, we use a dynamic scheduling scheme for assigning points to threads since the neighborhood sizes of objects may vary. Obviously, this forms an almost perfect target for parallelizing PreDeCon (though Ideal is not a clustering algorithm and thus does not produce any clustering result). We then use this algorithm for studying the performance of AnyPDC-MC on the real datasets ED, BCI, and WT. The results are shown in Fig. 15.

As we can see in Fig. 15 (top), AnyPDC-MC acquires very good speedup factors for all datasets and its performance is very close to those of the ideal algorithm. For the dataset ED, AnyPDC-MC and Ideal almost have the same performance. Since the neighborhood sizes of points vary significantly on the datasets BCI and WT, AnyPDC-MC performs slightly worse than Ideal due to the workload imbalance problem. However, when considering the final runtimes, AnyPDC-MC is usually faster than Ideal (and thus PreDeCon)[5] for most cases even though it has lower speedup factors with respect to the number of threads. Taking the dataset WT with $\epsilon = 0.06$ as an example, Ideal needs 102.7 s to finish and AnyPDC-MC ends after 32.0 s (3.2 times faster), while their speedup factors are 12.14 and 8.95, respectively. The reason is simply that AnyPDC-MC not only tries to enhance throughput over multiple threads, but also reduces the overall workload as shown in Sect. 4.1 (Fig. 12).

5   Related Work

Subspace Clustering. While traditional clustering algorithms focus on finding clusters in the full dimensionality of data, subspace clustering algorithms aim at finding clusters in subspace projections of high-dimensional datasets [31]. They have potential applications in many fields such as recommender systems [57,59–61,63], crowdsourcing [17,21,23–25,27], data integration [8,49,57], Web mining [9,10,19,55,62], data stream processing [14–16,18,20,48], information retrieval [26,54,56], and large-scale networks [22,28].

There are many different approaches for subspace clustering. For example, CLIQUE [4] divides the data space using grids and connects dense cells containing a significant number of points to build clusters. SUBCLU [33] extends the density-based clustering definition of DBSCAN [11] to subspace projections. PROCLUS [3] follows a k-medoid-like clustering scheme for producing non-overlapping clusters. Providing a complete survey of these techniques is beyond the scope of this paper. Interested readers please refer to other surveys for more information, e.g., [31,47,52].

PreDeCon and its Variants. Similar to PROCLUS, PreDeCon also produces non-overlapping, but density-based clusters using a specialized distance measure that captures the subspace of each cluster. However, it adapts the density-based cluster notions introduced by DBSCAN [11]. It can detect clusters of arbitrary shapes and is able to detect noise.

There exist some extensions of PreDeCon in the literature. IncPreDeCon [30] is an incremental version of PreDeCon for dynamic data. It relies on the locality of clustering structure changes for efficiently updating the results wrt. each inserted or deleted point in a database. HiSC [1] is a hierarchical extension

---

[5] Since Ideal ignores the cluster expansion process of PreDeCon, its runtime is obviously lower than that of PreDeCon itself.

of PreDeCon which produces a reachability plot containing an ordering of points and subspace distances, which is calculated based on the intersection of subspace preference vectors of two points p and q. Clusters can be built by cutting through the reachability plot with an arbitrary value of ϵ. Each separated part then forms a cluster. An extension of HiSC, DiSH [2], allows any cluster to be embedded in many other super-clusters thus leading to complex hierarchical relationship among clusters. HDDSTREAM [46] is another extension of PreDeCon designed to cope with streaming data. None of these algorithms is an anytime or parallel algorithm like AnyPDC. Also, none of them focuses on reducing the number of neighborhood queries for enhancing the performance of PreDeCon.

Anytime Algorithm. In [41], the authors introduce an anytime algorithm, called AnyDBC, for speeding up the density-based clustering method DBSCAN by reducing the number of range queries. The algorithm AnyPDC is actually built upon the concept of [41]. It first summarizes points into groups and tries to connect these groups together to form clusters by actively choosing points for processing until a termination condition is reached. However, their are several major differences between AnyDBC and AnyPDC. First, AnyPDC is designed to discover clusters hiding in subspaces of the data while AnyDBC is designed to find clusters in full dimensional space of the data. Second, to do so, AnyPDC relies on the preference weighted distance which does not satisfy the metric properties, which is mandatory for AnyDBC to work. Thus directly applying Any-DBC to the concept of subspace clustering is not possible. Therefore, AnyPDC must construct the cluster structure using a less strict condition. It exploits the lower-bounding distance of the preference weighted distance as a metric to determine the cluster structure while still guaranteeing the same final results as PreDeCon. Third, the cluster notions of AnyPDC and AnyDBC are different due to the constraint on the preference dimensionality of each point. Thus all the concepts of AnyDBC have to be extended to cope with this constraint.

6   Conclusion and Discussion

In this paper, we introduce for the first time an anytime extension of the fundamental clustering algorithm PreDeCon, called AnyPDC. At the same time, we show how to transform AnyPDC into a scalable parallel algorithm on multicore processors, thus uniquely making AnyPDC both a parallel and an interactive algorithm at the same time. To summarize, the major ideas of AnyPDC are (1) processing queries in block, (2) maintaining and updating an underlying graph structure representing local clusters, and (3) actively studying the current cluster structure and choosing only meaningful objects for examining their neighbors. Experiments on various real datasets show very interesting and promising results. AnyPDC approximates the final clustering results of PreDeCon very well at intermediate iterations. Even if it runs to the end, it is also faster than PreDeCon. Moreover, it scales very well with the number of threads in a shared memory environment.

Acknowledgments. We special thank to anonymous reviewers for their helpful comments. Part of this research was funded by a Villum postdoc fellowship, Vietnam National Foundation for Science and Technology Development (NAFOSTED) under grant number 102.05-2015.10 and the CDP Life Project.

References

1. Achtert, E., Böhm, C., Kriegel, H.-P., Kröger, P., Müller-Gorman, I., Zimek, A.: Finding hierarchies of subspace clusters. In: Fürnkranz, J., Scheffer, T., Spiliopoulou, M. (eds.) PKDD 2006. LNCS (LNAI), vol. 4213, pp. 446–453. Springer, Heidelberg (2006). https://doi.org/10.1007/11871637_42
2. Achtert, E., Böhm, C., Kriegel, H.-P., Kröger, P., Müller-Gorman, I., Zimek, A.: Detection and visualization of subspace cluster hierarchies. In: Kotagiri, R., Krishna, P.R., Mohania, M., Nantajeewarawat, E. (eds.) DASFAA 2007. LNCS, vol. 4443, pp. 152–163. Springer, Heidelberg (2007). https://doi.org/10.1007/978-3-540-71703-4_15
3. Aggarwal, C.C., Procopiuc, C.M., Wolf, J.L., Yu, P.S., Park, J.S.: Fast algorithms for projected clustering. In: SIGMOD, pp. 61–72 (1999)
4. Agrawal, R., Gehrke, J., Gunopulos, D., Raghavan, P.: Automatic subspace clustering of high dimensional data for data mining applications. In: SIGMOD, pp. 94–105 (1998)
5. Assent, I., Kranen, P., Baldauf, C., Seidl, T.: AnyOut: anytime outlier detection on streaming data. In: Lee, S., Peng, Z., Zhou, X., Moon, Y.-S., Unland, R., Yoo, J. (eds.) DASFAA 2012. LNCS, vol. 7238, pp. 228–242. Springer, Heidelberg (2012). https://doi.org/10.1007/978-3-642-29038-1_18
6. Böhm, C., Kailing, K., Kriegel, H.P., Kröger, P.: Density connected clustering with local subspace preferences. In: ICDM, pp. 27–34 (2004)
7. Chapman, B., Jost, G., Pas, R.: Using OpenMP: Portable Shared Memory Parallel Programming (Scientific and Engineering Computation). The MIT Press, Cambridge (2007)
8. Dang, M.T., Luong, A.V., Vu, T.-T., Nguyen, Q.V.H., Nguyen, T.T., Stantic, B.: An ensemble system with random projection and dynamic ensemble selection. In: Nguyen, N.T., Hoang, D.H., Hong, T.-P., Pham, H., Trawiński, B. (eds.) ACIIDS 2018. LNCS (LNAI), vol. 10751, pp. 576–586. Springer, Cham (2018). https://doi.org/10.1007/978-3-319-75417-8_54
9. Deng, X., Dou, Y., Lv, T., Nguyen, Q.V.H.: A novel centrality cascading based edge parameter evaluation method for robust influence maximization. IEEE Access **5**, 22119–22131 (2017)
10. Duong, C.T., Nguyen, Q.V.H., Wang, S., Stantic, B.: Provenance-based rumor detection. In: Huang, Z., Xiao, X., Cao, X. (eds.) ADC 2017. LNCS, vol. 10538, pp. 125–137. Springer, Cham (2017). https://doi.org/10.1007/978-3-319-68155-9_10
11. Ester, M., Kriegel, H.P., Sander, J., Xu, X.: A density-based algorithm for discovering clusters in large spatial databases with noise. In: KDD, pp. 226–231 (1996)
12. Greiner, J.: A comparison of parallel algorithms for connected components. In: SPAA, pp. 16–25 (1994)
13. Hinneburg, A., Aggarwal, C.C., Keim, D.A.: What is the nearest neighbor in high dimensional spaces? In: VLDB, pp. 506–515 (2000)

14. Hung, N.Q.V., Anh, D.T.: Combining sax and piecewise linear approximation to improve similarity search on financial time series. In: ISITC, pp. 58–62 (2007)
15. Hung, N.Q.V., Anh, D.T.: An improvement of PAA for dimensionality reduction in large time series databases. In: Ho, T.-B., Zhou, Z.-H. (eds.) PRICAI 2008. LNCS (LNAI), vol. 5351, pp. 698–707. Springer, Heidelberg (2008). https://doi.org/10.1007/978-3-540-89197-0_64
16. Hung, N.Q.V., Anh, D.T.: Using motif information to improve anytime time series classification. In: SoCPaR, pp. 1–6 (2013)
17. Hung, N.Q.V., et al.: Argument discovery via crowdsourcing. VLDB J. **26**, 511–535 (2017)
18. Hung, N.Q.V., Jeung, H., Aberer, K.: An evaluation of model-based approaches to sensor data compression. TKDE **25**, 2434–2447 (2013)
19. Hung, N.Q.V., Luong, X.H., Miklós, Z., Quan, T.T., Aberer, K.: An MAS negotiation support tool for schema matching. In: AAMAS, pp. 1391–1392 (2013)
20. Hung, N.Q.V., Sathe, S., Duong, C.T., Aberer, K.: Towards enabling probabilistic databases for participatory sensing. In: CollaborateCom, pp. 114–123 (2014)
21. Quoc Viet Hung, N., Tam, N.T., Tran, L.N., Aberer, K.: An evaluation of aggregation techniques in crowdsourcing. In: Lin, X., Manolopoulos, Y., Srivastava, D., Huang, G. (eds.) WISE 2013. LNCS, vol. 8181, pp. 1–15. Springer, Heidelberg (2013). https://doi.org/10.1007/978-3-642-41154-0_1
22. Hung, N.Q.V., Tam, N.T., Miklós, Z., Aberer, K.: On leveraging crowdsourcing techniques for schema matching networks. In: Meng, W., Feng, L., Bressan, S., Winiwarter, W., Song, W. (eds.) DASFAA 2013. LNCS, vol. 7826, pp. 139–154. Springer, Heidelberg (2013). https://doi.org/10.1007/978-3-642-37450-0_10
23. Hung, N.Q.V., Tam, N.T., Miklós, Z., Aberer, K.: Reconciling schema matching networks through crowdsourcing. EAI **1**, e2 (2014)
24. Hung, N.Q.V., et al.: Answer validation for generic crowdsourcing tasks with minimal efforts. VLDB J. **26**, 855–880 (2017)
25. Hung, N.Q.V., Thang, D.C., Weidlich, M., Aberer, K.: Minimizing efforts in validating crowd answers. In: SIGMOD, pp. 999–1014 (2015)
26. Nguyen, Q.V.H., Do, S.T., Nguyen, T.T., Aberer, K.: Tag-based paper retrieval: minimizing user effort with diversity awareness. In: Renz, M., Shahabi, C., Zhou, X., Cheema, M.A. (eds.) DASFAA 2015. LNCS, vol. 9049, pp. 510–528. Springer, Cham (2015). https://doi.org/10.1007/978-3-319-18120-2_30
27. Hung, N.Q.V., Viet, H.H., Tam, N.T., Weidlich, M., Yin, H., Zhou, X.: Computing crowd consensus with partial agreement. IEEE Trans. Knowl. Data Eng. **30**(1), 1–14 (2018)
28. Quoc Viet Nguyen, H., et al.: Minimizing human effort in reconciling match networks. In: Ng, W., Storey, V.C., Trujillo, J.C. (eds.) ER 2013. LNCS, vol. 8217, pp. 212–226. Springer, Heidelberg (2013). https://doi.org/10.1007/978-3-642-41924-9_19
29. Kleinberg, R.D.: Anytime algorithms for multi-armed bandit problems. In: SODA, pp. 928–936 (2006)
30. Kriegel, H.-P., Kröger, P., Ntoutsi, I., Zimek, A.: Density based subspace clustering over dynamic data. In: Bayard Cushing, J., French, J., Bowers, S. (eds.) SSDBM 2011. LNCS, vol. 6809, pp. 387–404. Springer, Heidelberg (2011). https://doi.org/10.1007/978-3-642-22351-8_24
31. Kriegel, H.P., Kröger, P., Zimek, A.: Clustering high-dimensional data: a survey on subspace clustering, pattern-based clustering, and correlation clustering. TKDD **3**(1), 1 (2009)

32. Kristensen, J., Mai, S.T., Assent, I., Jacobsen, J., Vo, B., Le, A.: Interactive exploration of subspace clusters for high dimensional data. In: Benslimane, D., Damiani, E., Grosky, W.I., Hameurlain, A., Sheth, A., Wagner, R.R. (eds.) DEXA 2017. LNCS, vol. 10438, pp. 327–342. Springer, Cham (2017). https://doi.org/10.1007/978-3-319-64468-4_25

33. Kröger, P., Kriegel, H.P., Kailing, K.: Density-connected subspace clustering for high-dimensional data. In: SDM, pp. 246–256 (2004)

34. Kumar, V.: Introduction to Parallel Computing, 2nd edn. Addison-Wesley Longman Publishing Co., Inc., Boston (2002)

35. Kywe, W.W., Fujiwara, D., Murakami, K.: Scheduling of image processing using anytime algorithm for real-time system. In: ICPR, vol. 3, pp. 1095–1098 (2006)

36. Mai, S.T., et al.: Scalable interactive dynamic graph clustering on multicore CPUs. TKDE

37. Mai, S.T., Amer-Yahia, S., Chouakria, A.D.: Scalable active temporal constrained clustering. In: EDBT, pp. 449–452 (2018)

38. Mai, S.T., Amer-Yahia, S., Chouakria, A.D., Nguyen, K.T., Nguyen, A.-D.: Scalable active constrained clustering for temporal data. In: Pei, J., Manolopoulos, Y., Sadiq, S., Li, J. (eds.) DASFAA 2018. LNCS, vol. 10827, pp. 566–582. Springer, Cham (2018). https://doi.org/10.1007/978-3-319-91452-7_37

39. Mai, S.T., Assent, I., Jacobsen, J., Dieu, M.S.: Anytime parallel density-based clustering. Data Min. Knowl. Discov. 32(4), 1121–1176 (2018)

40. Mai, S.T., Assent, I., Le, A.: Anytime OPTICS: an efficient approach for hierarchical density-based clustering. In: Navathe, S.B., Wu, W., Shekhar, S., Du, X., Wang, X.S., Xiong, H. (eds.) DASFAA 2016. LNCS, vol. 9642, pp. 164–179. Springer, Cham (2016). https://doi.org/10.1007/978-3-319-32025-0_11

41. Mai, S.T., Assent, I., Storgaard, M.: AnyDBC: an efficient anytime density-based clustering algorithm for very large complex datasets. In: SIGKDD, pp. 1025–1034 (2016)

42. Mai, S.T., Dieu, M.S., Assent, I., Jacobsen, J., Kristensen, J., Birk, M.: Scalable and interactive graph clustering algorithm on multicore CPUs. In: ICDE, pp. 349–360 (2017)

43. Mai, S.T., He, X., Feng, J., Böhm, C.: Efficient anytime density-based clustering. In: SDM, pp. 112–120 (2013)

44. Mai, S.T., He, X., Feng, J., Plant, C., Böhm, C.: Anytime density-based clustering of complex data. Knowl. Inf. Syst. 45(2), 319–355 (2015)

45. Mai, S.T., He, X., Hubig, N., Plant, C., Böhm, C.: Active density-based clustering. In: ICDM, pp. 508–517 (2013)

46. Ntoutsi, I., Zimek, A., Palpanas, T., Kröger, P., Kriegel, H.: Density-based projected clustering over high dimensional data streams. In: SDM, pp. 987–998 (2012)

47. Parsons, L., Haque, E., Liu, H.: Subspace clustering for high dimensional data: a review. SIGKDD Explor. 6(1), 90–105 (2004)

48. Peixoto, D.A., Hung, N.Q.V.: Scalable and fast top-k most similar trajectories search using mapreduce in-memory. In: Cheema, M.A., Zhang, W., Chang, L. (eds.) ADC 2016. LNCS, vol. 9877, pp. 228–241. Springer, Cham (2016). https://doi.org/10.1007/978-3-319-46922-5_18

49. Peixoto, D.A., Zhou, X., Hung, N.Q.V., He, D., Stantic, B.: A system for spatial-temporal trajectory data integration and representation. In: Pei, J., Manolopoulos, Y., Sadiq, S., Li, J. (eds.) DASFAA 2018. LNCS, vol. 10828, pp. 807–812. Springer, Cham (2018). https://doi.org/10.1007/978-3-319-91458-9_53

50. Settles, B.: Active learning literature survey. Computer Sciences Technical report 1648, University of Wisconsin-Madison (2009)

51. Shieh, J., Keogh, E.J.: Polishing the right apple: anytime classification also benefits data streams with constant arrival times. In: ICDM, pp. 461–470 (2010)
52. Sim, K., Gopalkrishnan, V., Zimek, A., Cong, G.: A survey on enhanced subspace clustering. Data Min. Knowl. Discov. **26**(2), 332–397 (2013)
53. Smyth, P., Wolpert, D.: Anytime exploratory data analysis for massive data sets. In: KDD, pp. 54–60 (1997)
54. Tam, N.T., Hung, N.Q.V., Weidlich, M., Aberer, K.: Result selection and summarization for web table search. In: ICDE, pp. 231–242 (2015)
55. Tam, N.T., Weidlich, M., Thang, D.C., Yin, H., Hung, N.Q.V.: Retaining data from streams of social platforms with minimal regret. In: IJCAI, pp. 2850–2856 (2017)
56. Thang, D.C., Tam, N.T., Hung, N.Q.V., Aberer, K.: An evaluation of diversification techniques. In: Chen, Q., Hameurlain, A., Toumani, F., Wagner, R., Decker, H. (eds.) DEXA 2015. LNCS, vol. 9262, pp. 215–231. Springer, Cham (2015). https://doi.org/10.1007/978-3-319-22852-5_19
57. Toan, N.T., Cong, P.T., Tam, N.T., Hung, N.Q.V., Stantic, B.: Diversifying group recommendation. IEEE Access **6**, 17776–17786 (2018)
58. Ueno, K., Xi, X., Keogh, E.J., Lee, D.J.: Anytime classification using the nearest neighbor algorithm with applications to stream mining. In: ICDM, pp. 623–632 (2006)
59. Wang, W., Yin, H., Huang, Z., Sun, X., Hung, N.Q.V.: Restricted Boltzmann machine based active learning for sparse recommendation. In: Pei, J., Manolopoulos, Y., Sadiq, S., Li, J. (eds.) DASFAA 2018. LNCS, vol. 10827, pp. 100–115. Springer, Cham (2018). https://doi.org/10.1007/978-3-319-91452-7_7
60. Yin, H., Chen, H., Sun, X., Wang, H., Wang, Y., Nguyen, Q.V.H.: SPTF: a scalable probabilistic tensor factorization model for semantic-aware behavior prediction. In: ICDM, pp. 585–594 (2017)
61. Yin, H., Chen, L., Wang, W., Du, X., Hung, N.Q.V., Zhou, X.: Mobi-SAGE: a sparse additive generative model for mobile app recommendation. In: ICDE, pp. 75–78 (2017)
62. Yin, H., et al.: Discovering interpretable geo-social communities for user behavior prediction. In: ICDE, pp. 942–953 (2016)
63. Yin, H., Zhou, X., Cui, B., Wang, H., Zheng, K., Hung, N.Q.V.: Adapting to user interest drift for POI recommendation. TKDE **28**, 2566–2581 (2016)
64. Zaki, M.J., Meira Jr., W.: Data Mining and Analysis: Fundamental Concepts and Algorithms. Cambridge University Press, New York (2014)
65. Zilberstein, S.: Using anytime algorithms in intelligent systems. AI Mag. **17**(3), 73–83 (1996)
66. Zilberstein, S., Russell, S.J.: Anytime sensing planning and action: a practical model for robot control. In: IJCAI, pp. 1402–1407 (1993)

MapFIM+: Memory Aware Parallelized Frequent Itemset Mining In Very Large Datasets

Khanh-Chuong Duong[1,2], Mostafa Bamha[2(✉)], Arnaud Giacometti[1],
Dominique Li[1], Arnaud Soulet[1], and Christel Vrain[2]

[1] Université de Tours, LIFAT EA 6300, Blois, France
`{arnaud.giacometti,dominique.li,arnaud.soulet}@univ-tours.fr`
[2] Université d'Orléans, INSA Centre Val de Loire, LIFO EA 4022, Orléans, France
`{khanh-chuong.duong,mostafa.bamha,christel.vrain}@univ-orleans.fr`

Abstract. Mining frequent itemsets in large datasets has received much attention in recent years relying on MapReduce programming model. For instance, many efficient Frequent Itemset Mining (a.k.a. FIM) algorithms have been parallelized to MapReduce principle such as *Parallel Apriori*, *Parallel FP-Growth* and *Dist-Eclat*. However, most approaches focus on job partitioning and/or load balancing without considering the extensibility depending on required memory assumptions. Thus, a challenge in designing parallel FIM algorithms consists therefore in finding ways to guarantee that data structures used during the mining process always fit in the local memory of processing nodes during all computation steps. In this paper, we propose MapFIM+, a two-phase approach to frequent itemset mining in very large datasets benefiting both from a MapReduce-based distributed Apriori method and local in-memory FIM methods. In our approach, MapReduce is first used to generate frequent itemsets until getting local memory-fitted prefix-projected databases, and an optimized local in-memory mining process is then launched to generate all remaining frequent itemsets from each prefix-projected database on individual processing nodes. Indeed, MapFIM+ improves our previous algorithm MapFIM by using an exact evaluation of prefix-projected database sizes during the MapReduce phase. This improvement makes MapFIM+ more efficient, especially for databases leading to huge candidate sets, by significantly reducing communication and disk I/O costs. Performance evaluation shows that MapFIM+ is more efficient and more extensible than existing MapReduce based frequent itemset mining approaches.

Keywords: Frequent itemset mining
MapReduce programming model
Distributed file systems · Hadoop framework

A. Hameurlain et al. (Eds.): TLDKS XXXIX, LNCS 11310, pp. 200–225, 2018.
https://doi.org/10.1007/978-3-662-58415-6_7

1   Introduction

Frequent pattern mining [2] is an important field of Knowledge Discovery in Databases that aims at extracting itemsets occurring frequently inside database entries (as transactions, event sets, etc.). Usually, a minimum support threshold is fixed in this problem and frequent patterns are defined as patterns whose frequency is greater than this threshold. All the algorithms rely on an important anti-monotonicity property for pruning the search space stating that when an itemset is extended then its support, i.e., the number of transactions it covers, decreases. In other terms, given an itemset, the supports of its supersets are lower or equal to its support. For more than 20 years, a large number of algorithms have been proposed to mine frequent patterns as efficiently as possible [1]. In the Big Data era, proposing efficient algorithms that handle huge volumes of transactions still remains an important challenge due to the memory space requirements while mining all frequent patterns. To tackle this issue, recent approaches have been made to work in distributed environments and the major idea is to distinguish two mining phases: a global phase and a local one. In such schemes, the first global mining phase often relies on MapReduce [6] to find among the frequent patterns the ones whose calculation requires a huge amount of data that does not fit in memory; then, the local mining phase mines on single nodes all the supersets of the patterns obtained at the global phase. Obviously, the idea is that such supersets can be extracted using a part of data that can fit in the memory of a single machine. Intuitively, the first phase guarantees the possibility of working on huge datasets while the second phase preserves a reasonable execution time. Unfortunately, existing approaches are all difficult to be fully extensible, *i.e.* mining becomes intractable as soon as the number of transactions is too large or the minimum frequency threshold is too low.

For the efficiency of two-phase frequent mining approaches, a major difficulty consists in determining the balance between the global and the local mining phases. Indeed, if an approach relies too heavily on local mining, it will be limited to large minimum frequency thresholds only, for which the amount of candidate patterns and/or the projected databases fit in memory. For instance, Parallel FPF algorithm [8] with distributed projected databases cannot deal with very low minimum thresholds when projected databases do not fit in local memory of a machine. Conversely, if an approach relies too much on global mining, it will be less efficient since the cost of communications is high. For instance, Parallel Apriori [9] is relatively slow for low thresholds because all patterns are extracted during the global phase. In BigFIM [15], a parameter k must be set by the users: it represents the minimum length below which the itemsets are mined globally while itemsets that are larger than k are locally mined since they cover a smaller set of transactions that is supposed to fit in memory. However, a practical problem is that such a length is difficult to determine as it depends on datasets and on available memory. To illustrate the issue raised by the setting of the threshold k, Fig. 1 plots the maximum lengths of frequent itemsets with the WebDocs dataset (see Sect. 5 for details) when the minimum frequency threshold varies. In [15], it is suggested to use a global phase for itemsets of size $k = 3$, assuming that, for larger itemsets, the conditional databases will fit in memory. However, from Fig. 1, with respect to the used dataset, it is easy to see that 3 is not a sufficiently high threshold since there is at least one itemset of size 4 that covers more than 40% of transactions. Moreover, two patterns of the same size may have very different frequencies and this point is not taken into account in BigFIM.

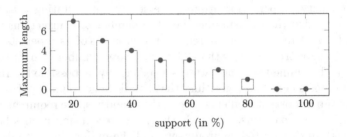

Fig. 1. Maximum length of frequent itemsets in WebDocs dataset

We introduced in [7] MapFIM (Memory aware parallelized Frequent Itemset Mining) algorithm which is, to the best of our knowledge, the first algorithm extensible with respect to the number of transactions. The advantage of this extensibility is that, it is possible to process large volumes of data (although the addition of machines does not necessarily improve run-time performance as it is the case with scalability). The key idea is to introduce a maximum frequency threshold β above which frequency counting for an itemset is distributed on several machines. We proved that there exists at least one setting of β for which the algorithm is extensible under the conditions that the FIM algorithm used locally takes a memory space bounded with respect to the size of a projected database and that the set of items holds in memory. We showed how to determine this parameter in practice. Indeed, the higher this threshold, the faster the mining (because more patterns are mined locally). Performance evaluation showed the extensibility and the efficiency of MapFIM compared to the best state-of-the-art algorithms. Nevertheless, in MapFIM, β parameter is estimated using the average of transaction lengths. This may induce a rough estimation of β parameter in some extreme database cases, whereas β is the criterion used to switch from the global to the local mining phase, and finding an accurate evaluation of β is fundamental for efficiency.

In this paper we introduce MapFIM+, an extended version of MapFIM. Map-FIM+ improves our previous approach by using an efficient MapReduce routine to generate candidate sets, which makes it more efficient than MapFIM by reducing communication and disks I/O costs; this is all the more true in the case of huge candidate sets. Moreover, MapFIM+ is based on an exact evaluation of prefix-projected database sizes which makes MapFIM+ insensitive to an estimation error of MapFIM's β parameter. However, to be able to adapt MapFIM+ to the amount of memory available on processing nodes for local mining phase, we introduce a simplified parameter called γ that is used as a prefix-projected database size threshold for local mining. This parameter is fixed only using the amount of memory available on processing nodes.

Contributions

- We present a transaction-extensible algorithm MapFIM+ for mining frequent itemsets, i.e., it manages to mine all frequent itemsets whatever the number of transactions.
- MapFIM+ is an extension of MapFIM, that relies on a single parameter γ, depending only on the amount of memory available on processing nodes. The parameter γ allows to determine when the frequent supersets of a frequent itemset can be mined locally.
- We prove that our algorithm is correct and complete under the condition that the local mining algorithm is prefix-complete. We also prove that our algorithm is transaction-extensible.
- We propose a method for automatically calibrating γ.
- We conduct experiments, on WebDocs dataset and on artificially generated datasets, allowing to compare MapFIM+ first with MapFIM and then, with the best approaches for itemset mining using Hadoop MapReduce framework (BigFIM and PFP). We also illustrate the transaction-extensibility of Map-FIM+, showing that it can deal with databases that BigFIM and PFP can not handle.

The rest of the paper is organized as follows. Section 2 formulates the problem of frequent itemset mining in an extensible way in order to deal with a huge volume of transactions. Section 3 studies existing approaches in literature in terms of extensibility. In Sect. 4, we present how our improved algorithm MapFIM+ works and in particular, we detail the two phases (i.e., global and local mining processes) and the switch between them. In Sect. 5, we empirically compare the performance of MapFIM+ to MapFIM [7] and also against the state-of-the-art methods by comparing execution times and memory consumption. Section 6 concludes this paper.

2   Problem Formulation

2.1   Frequent Itemset Mining Problem

Let $\mathcal{I} = \{i_1, i_2, \ldots, i_n\}$ be a set of n literals called *items*. An itemset (or a pattern) is a subset of $\mathcal{I}$. The language of itemsets corresponds to $2^{\mathcal{I}}$. A transactional database $\mathcal{D} = \{t_1, t_2, \ldots, t_m\}$ is a multi-set of itemsets of $2^{\mathcal{I}}$. Each itemset t_i, usually called a *transaction*, is a database entry. For instance, Table 1 gives a transactional database with 10 transactions t_i described by 6 items $\mathcal{I} = \{a, b, c, d, e, f\}$.

Table 1. Original dataset

Transaction	items	Transaction	items
t_1	a	t_6	a, d
t_2	a, b	t_7	b, c
t_3	a, b, c	t_8	c, d
t_4	a, b, c, d	t_9	c, e
t_5	a, c	t_{10}	f

Pattern discovery takes advantage of interestingness measures to evaluate the relevancy of an itemset. The *frequency* of an itemset X in the transactional database $\mathcal{D}$ is the number of transactions covered by X [2]: $freq(X, \mathcal{D}) = |\{t \in \mathcal{D} : X \subseteq t\}|$ (or $freq(X)$ for sake of brevity). Then, the *support* of X is its proportion of covered transactions in $\mathcal{D}$: $supp(X, \mathcal{D}) = freq(X, \mathcal{D})/|\mathcal{D}|$. An itemset is said to be *frequent* when its support exceeds a user-specified minimum threshold α. **Given a set of items $\mathcal{I}$, a transactional database $\mathcal{D}$ and a minimum support threshold, frequent itemset mining (FIM) aims at enumerating all frequent itemsets.**

In our approach, in order to distribute the computation of frequent itemsets, we use the usual notion of *projected database*. More precisely, given an arbitrary total order over the set of all items, the projected database of an itemset is defined as follows:

Definition 1 (Projected database). *Given an arbitrary total order $<_\mathcal{I}$ over the set $\mathcal{I}$ of all items and a database $\mathcal{D}$, let X be an itemset. The projected database of X, denoted $\mathcal{D}_X$, is defined by: $\mathcal{D}_X = \{\sigma_{>X}(Y) : Y \in \mathcal{D}, X \subset Y\}$ where $\sigma_{>X}(Y) = \{i \in Y : (\forall j \in X)(j <_\mathcal{I} i)\}$. Moreover, the size of $\mathcal{D}_X$, denoted $\|\mathcal{D}_X\|$, is defined by $\|\mathcal{D}_X\| = \sum_{Y \in \mathcal{D}_X} |Y|$.*

For instance, considering the database $\mathcal{D}$ presented Table 1 and the total order $a <_\mathcal{I} b <_\mathcal{I} \cdots <_\mathcal{I} f$, the projected database of itemset ab is $\mathcal{D}_{ab} = \{c, cd\}$ and we have $\|\mathcal{D}_{ab}\| = 1 + 2 = 3$.

2.2   MapReduce Programming Model

MapReduce is a simple yet powerful programming model initialized by Google [6] for implementing distributed applications without having extensive prior knowledge of issues related to data redistribution, task allocation or fault tolerance in large scale distributed systems.

The core functioning of MapReduce is based on two functions, **map** and **reduce**, that developers are supposed to provide to the framework. These two functions should have the following signatures:

Map: $(k_1, v_1) \longrightarrow list(k_2, v_2),$
Reduce: $(k_2, list(v_2)) \longrightarrow list(k_3, v_3).$

The **map** function has two input parameters, a key k_1 and an associated value v_1, and outputs a list of intermediate key/value pairs (k_2, v_2). This list is partitioned by the MapReduce framework depending on the values of k_2, with the constraint that all elements with the same value of k_2 belong to the same group. The **reduce** function has two parameters as inputs: an intermediate key k_2 and a list of intermediate values $list(v_2)$ associated with k_2. It applies the user defined merge logic on $list(v_2)$ and outputs a list of values $list(k_3, v_3)$.

MapReduce excels in the treatment of data parallel applications, where computation can be decomposed into many independent tasks, involving large input data. However MapReduce's performance may degrade in the case of dependent tasks or in the presence of skewed data due to the fact that, in Map phase, all the emitted key-value pairs (k_2, v_2) corresponding to the same key k_2 are sent to the same reducer. This may induce a load imbalance among processing nodes and also can lead to task failures whenever the list of values corresponding to a specific key k_2 cannot fit in processing nodes available memory [3,4]. For scalability, MapReduce algorithm's design must avoid load imbalance among processing nodes while reducing disks I/O and communication costs during all stages of MapReduce jobs computation.

In this paper, our approach is based on Hadoop, the industrial standard open source implementation of MapReduce as well on its distributed file system HDFS (Hadoop Distributed File System) designed to store very large files with streaming data access patterns.

2.3 The Challenge of Extensibility

Guaranteeing the correct execution of a method whatever the volume of input data is a classical challenge in MapReduce through the notion of scalability. Scalability refers to the capacity of a method to perform similarly even if there is a change in the order of magnitude of the data volume, in particular by adding new machines (as mappers or reducers). We introduce the notion of *extensibility*, which refers to the capacity of a method to deal with an increase in the data volume but without performance guarantees.

More precisely, our goal is to efficiently process transaction databases whatever the number of transactions while the set of all distinct items remains unchanged. This situation covers many practical use cases. For instance, in a supermarket, the set of products is relatively stable while new transactions will be added continuously. We formalize the notion of extensibility with respect to the number of transactions as follows:

Definition 2 (Transaction-extensible). *Given a set of items $\mathcal{I}$, a FIM method is said to be transaction-extensible iff it manages to mine all frequent itemsets whatever the number of transactions in $\mathcal{D} = \{t_1, \ldots, t_m\}$ (where $t_i \subseteq \mathcal{I}$) and the minimum support threshold α.*

This definition is particularly interesting for a pattern discovery task. Indeed, the transaction-extensible property guarantees that for a given set of items $\mathcal{I}$, the method will always be able to mine all the frequent itemsets whatever the changes of the number of transactions in $\mathcal{D}$ and of the minimum frequent threshold α. In this paper, we aim at proposing the first transaction-extensible FIM method. This goal is clearly a challenge in terms of controlling the amount of memory required by frequent itemset mining.

3   Related Work

Due to the explosive growth of data, many parallel implementations of frequent pattern mining (FPM) algorithms have been proposed in the literature, mainly on mining frequent itemsets [8–10,16–18,21], but also to mine frequent sequences [5,14]. In this section, we only consider related work involving the parallelization of FPM algorithms via MapReduce and review their important shortcomings.

A first category of approaches includes approaches that are specific parallelizations of existing FPM algorithms, for example, different adaptations of Apriori on MapReduce [9,10]. These adaptations of Apriori are not *transaction-extensible* since they assume that at each level, the set of candidate itemsets can be stored in the main memory of the worker nodes (mappers or reducers). We show in Sect. 4.3 how this limitation can be overcome by using HDFS to store the set of candidates. Different implementations of FP-Growth on MapReduce [8,21] distribute the conditional databases of the frequent items to the mappers. However, these proposals do not guarantee that the conditional databases can be stored among worker nodes, and therefore, these parallelizations of FP-Growth are also not *transaction-extensible*. More recently, Makanju et al. [13] propose to use Parent-Child MapReduce (a new feature of IBM Platform Symphony) to overcome the limitations of the previous implementations of FP-Growth and show that their method provides significant speed-ups over Parallel FP-Growth [8]. However, their method requires to predict the processing loads of a FP-Tree which is a particularly difficult challenge.

A second category of approaches includes approaches that are independent of a specific FPM algorithm, meaning that after a data preparation and partitioning phase, they can use any existing FPM methods to locally extract patterns. In this category, we can distinguish two sub-categories of approaches as follows.

At a high-level, the methods in the first sub-category carefully partition the original dataset in such a way that each partition can be mined independently and in parallel [14,16,17]. In [18], the authors propose an algorithm that extract frequent itemsets in three phases. In the first phase, their algorithm divides the original dataset $\mathcal{D}$ into a number of non-overlapping partitions $\mathcal{D}_k$ ($k \in [1..K]$). Then, in the second phase, each partition $\mathcal{D}_k$ is mined independently to extract itemsets that are locally frequent, i.e. frequent in $\mathcal{D}_k$. Finally, in the third phase, the sets of locally frequent itemsets are merged, and a scan of the whole dataset is performed to identify the itemsets that are globally frequent, i.e. frequent in $\mathcal{D}$. Note that this third phase is necessary because the sets of locally frequent itemsets may contain false positive, i.e. itemsets that are locally frequent but not globally frequent. In order to overcome this problem and remove the need of the third phase, the algorithms proposed in [14, 16, 17] introduced partition methods such that locally frequent patterns are necessarily globally frequent. However, it is important to note that to achieve this, partitions $\mathcal{D}_k$ can overlap and some frequent patterns can be generated several times. Moreover, all these methods cannot guarantee that the data partitions $\mathcal{D}_k$ will fit in main memory (of the mappers or reducers), which means that they are not *transaction-extensible*.

The approaches in the second sub-category do not initially partition the dataset, but the search space (the pattern language), thereby ensuring that each frequent pattern is only generated once. We can consider that Parallel FP-Growth (PFP) [9] also belongs to this second sub-category of methods. However, because PFP partitions the search space only considering single frequent items, it is less efficient. In order to overcome this type of limitation, Moens et al. [15] propose in BigFIM to use longer frequent itemsets as prefixes for partitioning the search space. In a first and global phase, BigFIM mines the frequent k-itemsets using a MapReduce implementation of Apriori, and then subsets of prefixes of length k are passed to worker nodes in a second phase. These worker nodes use the conditional databases of prefixes to mine frequent patterns that are more specific, assuming that the conditional databases can fit in the main memory of the worker nodes.

Note that the selection of the parameter k for BigFIM can be very difficult in practice. Indeed, if a too low value is chosen for k, BigFIM might not terminate successfully if any conditional database cannot fit in main memory; on the other hand, if the k value is too high, the first global phase that computes the frequent k-itemsets will be highly time consuming. That is also why we propose our novel approach MapFIM+ that do not require any involvement to fix a parameter such as k, and automatically detect when it is possible to switch from global mining to local mining.

4   MapFIM+: An Improved MapReduce Approach for Frequent Itemset Mining

4.1   Overview of the Approach

The key idea of our proposal is to enumerate using a breadth-first search all itemsets using distributed techniques (global mining phase) until one reaches a point of the search space where all its supersets can be mined on a single machine (local mining phase). This point of the search space is reached as soon as each projected database (plus the amount of memory required to enumerate the itemsets) holds in memory. To do this, for each itemset, we do not only compute its frequency, but also the size of its projected database.

More precisely, in order to evaluate when it is possible to switch to the local mining phase, we introduce a maximum projected database threshold γ and define below the notions of *locally tractable* itemset and *minimally locally tractable* itemset.

Definition 3 (Locally tractable). *Given a database* $\mathcal{D}$, *a minimum support threshold* α *and a maximum projected database threshold* γ, *an itemset* X *is said to be* locally tractable *if it is frequent and its projected database holds in memory, i.e.* $supp(X, \mathcal{D}) \geq \alpha$ *and* $\|\mathcal{D}_X\| \leq \gamma$. *Moreover, an itemset* $X = (i_1, \dots, i_k)$ *with* $i_1 <_{\mathcal{I}} \dots <_{\mathcal{I}} i_k$, *is said to be* minimally locally tractable *if it is* locally tractable *and* $(i_1, \dots, i_{k-1})$ *is not* locally tractable.

In the following, we denote $\mathcal{T}^+$ *the set of frequent itemsets that are* minimally *locally tractable, e.g. itemsets* X *such that* $supp(X, \mathcal{D}) \geq \alpha$, $\|\mathcal{D}_X\| \leq \gamma$ *(locally tractable) and that are minimal among the* locally tractable *itemsets. We also denote* $\mathcal{T}^-$ *the set of frequent itemsets that are not* locally tractable, e.g. *itemsets* X *such that* $supp(X, \mathcal{D}) \geq \alpha$ *and* $\|\mathcal{D}_X\| > \gamma$. $\mathcal{G}$ *is the set of frequent itemsets that are either* not *locally tractable or* minimally *locally tractable* $(\mathcal{G} = \mathcal{T}^+ \cup \mathcal{T}^-)$. *Such sets are indexed by* k *when we want to refer to itemsets of length* k.

For instance, let us consider the database $\mathcal{D}$ presented in Table 1, the minimum support threshold $\alpha = 20\%$ and the maximum projected database threshold $\gamma = 5$. Note that itemset $\{a\}$ is not *locally tractable* as its projected database $\mathcal{D}_a = \{b, bc, bcd, c, d\}$ and $\|\mathcal{D}_a\| = 1 + 2 + 3 + 1 + 1 = 8 > \gamma$. On the other hand, itemset $\{ab\}$ is *locally tractable* as $freq(ab, \mathcal{D}) = 3 \geq \alpha.|\mathcal{D}| = 2$ (ab is frequent), $\mathcal{D}_{ab} = \{c, cd\}$ and $\|\mathcal{D}_{ab}\| = 1 + 2 = 3 < \gamma$ (the projected database of ab holds in memory). Moreover, itemset $\{ab\}$ is *minimally locally tractable* since itemset $\{a\}$ is not *locally tractable*.

Given a transactional database $\mathcal{D}$, a minimum support threshold α and a maximum projected data threshold γ, as shown in Algorithm 1, MapFIM+ enumerates all frequent itemsets by using three main phases:

1. **Initialization and database compression:** This phase initializes the process by compressing the original database $\mathcal{D}$ based on frequent 1-itemsets (see line 3 of Algorithm 1). Considering the database $\mathcal{D}$ presented in Table 1 and $\alpha = 20\%$, the compressed version of $\mathcal{D}$, denoted $\mathcal{D}'$, is shown in Table 2. $\mathcal{D}$ is first compressed by removing non-frequent items e and f. Then, transactions t_1 and t_{10} in $\mathcal{D}$ are removed because they can not contain an itemset of size 2 (or greater). Next, Algorithm 1 computes the set $\mathcal{T}_1^-$ of 1-itemsets in $\mathcal{G}_1$ that are not locally tractable (see line 4) and the set $\mathcal{T}_1^+$ of 1-itemsets that are *minimally locally tractable* (see line 5). In our example, considering $\alpha = 20\%$ and $\gamma = 5$, we obtain $\mathcal{G}_1 = \{a, b, c, d\}$. Note that itemset $\{a\}$ is not *locally tractable* as its projected database $\mathcal{D}'_a$ consists of 5 transactions (see Table 3) and $\|\mathcal{D}_a\| = 8 > \gamma$. Conversely, itemsets $\{b\}$, $\{c\}$ and $\{d\}$ are *locally tractable* since $\|D'_b\| = 4 < 5$, $\|D'_c\| = 2 < 5$ and $\|D'_d\| = 0 < 5$. Therefore, at the end of this phase, we have $\mathcal{T}_1^- = \{a\}$ and $\mathcal{T}_1^+ = \{b, c, d\}$.

2. **Global mining:** This phase mines all potentially frequent itemsets that are not *locally tractable* using Apriori algorithm. At each iteration k, Algorithm 1 first generates the set of candidate k-itemsets $\mathcal{C}_k$ by joining $\mathcal{T}_{k-1}^-$ and $\mathcal{G}_{k-1}$ (see line 9). Note that we do not join $\mathcal{G}_{k-1}$ with $\mathcal{G}_{k-1}$ because we do not want to generate candidates that are *locally tractable*, except candidates that are potentially *minimally locally tractable*. For instance, at step $k = 2$,

three candidates of size 2 will be generated from itemsets in $\mathcal{T}_1^- = \{a\}$ and $\mathcal{G}_1 = \{a, b, c, d\}$, e.g. $\mathcal{C}_2 = \{ab, ac, ad\}$. Indeed, these candidate 2-itemsets are potentially not *locally tractable*.

After the generation of the set of candidate k-itemsets $\mathcal{C}_k$, Algorithm 1 evaluates their frequency and the size of their projected database (see line 11). Using these measures, it computes the set $\mathcal{G}_k$ of frequent k-itemsets (mined during the global phase). Then, it identifies from $\mathcal{G}_k$ the set of frequent k-itemsets that are not *locally tractable*, e.g. the set $\mathcal{T}_k^-$ (see line 13). Finally, it computes the set $\mathcal{T}_k^+$ of frequent k-itemsets that are *minimally locally tractable*. We will demonstrate in Sect. 4.5 that all itemsets in $\mathcal{G}_k \setminus \mathcal{T}_k^-$ are necessarily *minimally locally tractable*. In our example, all candidate 2-itemsets in $\mathcal{C}_2 = \{ab, ac, ad\}$ are frequent. Indeed, we have: $freq(ab, \mathcal{D}') = 3 \geq \alpha.|\mathcal{D}| = 2$, $freq(ac, \mathcal{D}') = 3 \geq 2$ and $freq(ad, \mathcal{D}') = 2 \geq 2$. Moreover, because the size of their projected databases ($\|\mathcal{D}'_{ab}\| = 3$, $\|\mathcal{D}'_{ac}\| = 1$ and $\|\mathcal{D}'_{ad}\| = 0$) is lower than $\gamma = 5$, $\mathcal{T}_2^-$ is empty and $\mathcal{T}_2^+ = \{ab, ac, ad\}$. Finally, because $\mathcal{T}_2^- = \emptyset$, MapFIM+ moves to the *local mining* phase (see the test at line 7).

3. **Local mining:** This phase mines the itemsets from frequent itemsets that are *minimally locally tractable* (see line 17). In our running example, the frequency of the prefix-based supersets generated from $\mathcal{T}^+ = \mathcal{T}_1^+ \cup \mathcal{T}_2^+ = \{b, c, d, ab, ac, ad\}$ will be evaluated during this phase. Each prefix is considered individually by using a projected database as given in Table 3. More precisely, frequent pattern abc will be generated from the prefix ab, frequent pattern bc from the prefix b, frequent pattern cd from c, and we will obtain $\mathcal{L} = \{abc, bc, cd\}$. Finally, Algorithm 1 will return the set of all frequent patterns $\mathcal{G}_1 \cup \mathcal{G}_2 \cup \mathcal{L} = \{a, b, c, d\} \cup \{ab, ac, ad\} \cup \{abc, bc, cd\}$.

Table 2. Compressed database $\mathcal{D}'$	
Transaction	Items
t_2	a, b
t_3	a, b, c
t_4	a, b, c, d
t_5	a, c
t_6	a, d
t_7	b, c
t_8	c, d

Table 2. Compressed database $\mathcal{D}'$

Itemset	Projected databases
a	$\mathcal{D}'_a = \{b, bc, bcd, c, d\}$
b	$\mathcal{D}'_b = \{c, cd, c\}$
c	$\mathcal{D}'_c = \{d, d\}$
d	$\mathcal{D}'_d = \emptyset$
ab	$\mathcal{D}'_{ab} = \{c, cd\}$
ac	$\mathcal{D}'_{ac} = \{d\}$
ad	$\mathcal{D}'_{ad} = \emptyset$
cd	$\mathcal{D}'_{cd} = \emptyset$

Table 3. Projected databases

In the following, Sects. 4.2, 4.3, and 4.4 detail how MapReduce is used to implement efficiently the three main phases of MapFIM+. Then, Sect. 4.5 demonstrates the completeness and extensibility of MapFIM+ with respect to the number of transactions.

Algorithm 1. MapFIM+: chaining of MapReduce jobs

---

1 **Function** Main(*Float* α, *Float* γ):
2   // Initialization and database compression phase;
3   Computation of $\mathcal{G}_1$ and generation of compressed database $\mathcal{D}'$ from $\mathcal{D}$;
4      $\mathcal{T}_1^- \leftarrow \{X \in \mathcal{G}_1 : \|D'_X\| > \gamma\}; \mathcal{T}_1^+ \leftarrow \mathcal{G}_1 \setminus \mathcal{T}_1^-$ and $k \leftarrow 2$;
5   // Global Mining phase;
6   **while** $|\mathcal{T}_{k-1}^-| > 0$ **do**
7   // Using GenMap/GenReduce job;
8   Generation of $\mathcal{C}_k$ from $\mathcal{T}_{k-1}^-$ and $\mathcal{G}_{k-1}$;
9   // Using EvalMap/EvalReduce job;
10   Evaluation of $freq(X, \mathcal{D}')$ and $\|D'_X\|$ for all candidates $X \in \mathcal{C}_k$;
11         $\mathcal{G}_k \leftarrow \{X \in \mathcal{C}_k : freq(X, D') \geq \alpha.|\mathcal{D}|\}$;
12         $\mathcal{T}_k^- \leftarrow \{X \in \mathcal{G}_k : \|D'_X\| > \gamma\}$;
13         $\mathcal{T}_k^+ \leftarrow \mathcal{G}_k \setminus \mathcal{T}_k^-$;
14         $k \leftarrow k + 1$;
15   // Local Mining phase using LocalMap/LocalReduce job;
16      $\mathcal{T}^+ = \bigcup_{i=1}^{k-1} \mathcal{T}_i^+$;
17   Computation of $\mathcal{L} \leftarrow \bigcup_{X \in \mathcal{T}^+} \{Y : X \subseteq Y \wedge freq(Y, \mathcal{D}') \geq \alpha.|\mathcal{D}|\}$;
18   **return** $(\bigcup_{i=1}^{k-1} \mathcal{G}_i\}) \cup \mathcal{L}$;

4.2 Database Compression and Initialization

In this phase, we have first to compute the set of frequent 1-items, *e.g.* the set of items in $\mathcal{G}_1$. Using MapReduce, this goal can be achieved by adapting the *Word Count* routine [6]. Each item is considered as a word and we get the support of every item by MapReduce word counting. Then, the compressed database $\mathcal{D}'$ is generated and stored in HDFS. This procedure is solved by a simple Map function, where each mapper reads a block of data, removes items which are not in $\mathcal{G}_1$, and finally emits transactions with at least two frequent items.

Finally, we have to compute the size of the projected databases of all frequent items to determine if they are *locally tractable* or not (see line 4 of Algorithm 1). This goal can also be achieved using MapReduce. In the Map function, for each transaction $t \in \mathcal{D}'$, a pair (i, s) is emitted where i is an item in t and s is the length of t minus the position of i in t. In the Reduce function, for each pair (i, L) received, we just have to sum the values in L to obtain the size of the projected database $\mathcal{D}'_i$ of item i. For example, for transactions t_2, t_3, t_4 and t_7, the mappers will emit the pairs $(b, 0)$, $(b, 1)$, $(b, 2)$ and $(b, 1)$ and a reducer will compute the size of the projected database $\mathcal{D}'_b$ of item b as $0 + 1 + 2 + 1 = 4$. At the same time, using the γ parameter and the size of the projected database of all frequent items, the sets $\mathcal{T}_1^-$ and $\mathcal{T}_1^+$ can be easily constructed.

4.3 Global Mining Based on Apriori

This phase is similar to the parallel implementation of Apriori algorithm [9]. The key difference is mainly on the way candidates are generated (see Sect. 4.3).

Moreover, during the evaluation of the supports of the candidates (see the following Candidate Evaluation Step), we also evaluate the size of their projected databases, in order to detect whether they are *minimally locally tractable* or not.

Candidate Generation Step. In Apriori algorithm and its parallel implementation [9], the set $\mathcal{C}_{k+1}$ of candidate $(k + 1)$-itemsets is generated by the join $\mathcal{L}_k \bowtie \mathcal{L}_k$ at each iteration, where $\mathcal{L}_k$ denotes the set of all frequent k-itemsets[1].

In our case, a candidate $(k + 1)$-itemset is obtained by the join of a frequent but not *locally tractable* k-itemset, i.e. an itemset in $\mathcal{T}_k^-$, with a frequent k-itemset that is *locally tractable* or not, i.e. an itemset in $\mathcal{G}_k$. During the candidate evaluation step (see the following Candidate Evaluation Step), all candidate itemsets that are frequent are emitted and stored with a flag in $\{+, -\}$ to underline whether they are *locally tractable* $(flag = +)$ or not $(flag = -)$. Therefore, at iteration $(k + 1)$ of the candidate generation step, we join only frequent k-itemset with a negative flag $(flag = -)$ with frequent k-itemset (whatever their flag).

We now detail how we implement the candidate generation step using a MapReduce job (see Algorithm 2). In the Map function (see *GenMap* in Algorithm 2), for each pair $(X, flag) \in value$, where $X = (i_1, \ldots, i_{k-1}, i_k)$ is a frequent k-itemset and $flag \in \{+, -\}$ indicates whether X is *locally tractable* or not, we emit a pair $(prefix, (i_k, flag))$ where $prefix = (i_1, \ldots, i_{k-1})$ is the prefix of X of length $k - 1$.

In the Reduce function (see *GenReduce* in Algorithm 2), we combine candidate k-itemsets with the same prefix $P = (i_1, \ldots, i_{k-1})$. Given two pairs $(i, flag_i) \in values$ and $(j, flag_j) \in values$, we join the k-itemset $X = (i_1, \ldots, i_{k-1}, i)$ with the k-itemset $X' = (i_1, \ldots, i_{k-1}, j)$ if and only if:

- $flag_i = -$ in order to check whether X is not *locally tractable* (see line 14 of Algorithm 2), i.e. $X \in \mathcal{T}_k^-$ and
- $i < j$ in order to generate each candidate once (see line 16 of Algorithm 2).

If the two conditions are fullfilled, the reducer emits a new candidate $(k+1)$-itemset $Y = X \bowtie X' = (i_1, \ldots, i_{k-1}, i, j)$ (see line 17 of Algorithm 2). In Sect. 4.5, we prove that combining only frequent k-itemsets $X \in \mathcal{T}_k^-$ with frequent k-itemsets $X' \in \mathcal{G}_k$, we are complete w.r.t. $\mathcal{T}^+ \cup \mathcal{T}^-$.

Candidate Evaluation Step. The candidate evaluation step consists in computing both the frequency and the size of its projected database for each candidate in $\mathcal{C}_k$. This is achieved by a MapReduce job, described in Algorithm 3.

In the Map function (see *EvalMap* in Algorithm 3), for each transaction $t \in \mathcal{D}'$ and for each candidate $X \in \mathcal{C}_k$, if X is a subset of t (see line 8 in Algorithm 3), a pair composed of the value 1 (thus counting the presence of X in t) and the size of the projection of t in $\mathcal{D}'_X$ is emitted (see line 11 in

---

[1] In our work, in order to generate each candidate once, we use a prefix-based join operation. More precisely, given two sets of frequent k-itemsets $\mathcal{L}_k$ and $\mathcal{L}'_k$, the join of $\mathcal{L}_k$ and $\mathcal{L}'_k$ is defined by: $\mathcal{L}_k \bowtie \mathcal{L}'_k = \{(i_1, \ldots, i_k, i_{k+1}) \mid (i_1, \ldots, i_{k-1}, i_k) \in \mathcal{L}_k \wedge (i_1, \ldots, i_{k-1}, i_{k+1}) \in \mathcal{L}'_k \wedge i_1 < \cdots < i_k < i_{k+1}\}$.

Algorithm 2. MapFIM+: Map and Reduce functions to generate candidate itemsets

---

1 **Function** GenMap(*String key, String value*):
2   // *key*: input name;
3   // *value*: a set of pairs $(X, flag)$ where $X \in \mathcal{G}_k$ and $flag \in \{+, -\}$;
4   **foreach** $(X, flag) \in value$ **do**
5   Let $X = (i_1, \ldots, i_{k-1}, i_k)$;
6          $prefix \leftarrow (i_1, \ldots, i_{k-1})$;
7   Emit($prefix, (i_k, flag)$);

8 **Function** GenReduce(*String key, Iterator values*):
9   // *key*: a prefix $P = (i_1, \ldots, i_{k-1})$;
10   // *values*: a list of pairs $(i, flag)$ where $i \in \mathcal{I}$ and $flag \in \{+, -\}$;
11   **foreach** $(i, flag_i) \in values$ **do**
12   Let $X = (i_1, \ldots, i_{k-1}, i)$;
13   // Test if $X \in \mathcal{T}_k^-$;
14   **if** *(flag_i == −)* **then**
15   **foreach** $(j, flag_j) \in values$ **do**
16   **if** *(i < j)* **then**
17                     $Y \leftarrow (i_1, \ldots, i_{k-1}, i, j)$ // Y belongs to $\mathcal{C}_{k+1}$;
18   Emit($null, Y$) ;

Algorithm 3). More precisely, in order to compute the size of the projection of t in $\mathcal{D}'_X$, we first identify the maximal item i_{max} of X w.r.t. $<_\mathcal{I}$ (see line 9 in Algorithm 3). Then, we count the number of items j in t greater than i_{max} w.r.t. $<_\mathcal{I}$. Note that if the set of candidates $\mathcal{C}_k$ is too large to fit in memory of Mappers, then $\mathcal{C}_k$ is partitioned into blocks $\mathcal{C}_{Block}$ and Mappers process candidates block by block (see line 5 in Algorithm 3). Finally, we can point out that this Map phase achieves a good load balance because the compressed database $\mathcal{D}'$ is distributed equally among mappers and all mappers handle the same candidate set (reading the same number of candidate blocks).

In the Reduce function (see *EvalReduce* in Algorithm 3), each *key* is a candidate itemset $X \in \mathcal{C}_k$, and *value* is a list L including for each transaction $t \in \mathcal{D}'$ covered by X a pair $(freq, size)$ where $freq = 1$ and $size$ is the size of the projection of t in $\mathcal{D}'_X$. Thus, in order to compute the frequency of X and the size of $\mathcal{D}'_X$, a Reducer has just to sum the first and second components of the pairs in L (see lines 18 and 19). Then, if X is frequent in $\mathcal{D}$ w.r.t. the minimum support threshold α, i.e. $freq(X, \mathcal{D}') \geq \alpha \times |\mathcal{D}'|$ (see line 20 in Algorithm 3), a Reducer tests whether X is *locally tractable* or not (see line 21 in Algorithm 3). Finally, if X is *locally tractable*, we emit a pair $(X, +)$; otherwise, we emit a pair $(X, -)$ (see line 22 and 24 in Algorithm 3).

Algorithm 3. MapFIM+: Map and Reduce functions to evaluate candidates itemsets

---

```
 1  Function EvalMap(String key, String value):
 2    │ // key: input name;
 3    │ // value: a subset of transactions in D';
 4    │ while there are unsolved candidates do
 5    │   │ C_Block ← A block of C_k in HDFS;
 6    │   │ foreach transaction t ∈ value do
 7    │   │   │ foreach itemset X ∈ C_Block do
 8    │   │   │   │ if X ⊆ t then
 9    │   │   │   │   │ i_max ← max_<I (X) // Last item of X;
10    │   │   │   │   │ size ← |{j ∈ t : i_max <I j}|;
11    │   │   │   │   │ Emit(X, (1, size));

12  Function EvalReduce(String key, Iterator values):
13    │ // key: a candidate X ∈ C_k;
14    │ // values: a list of (frequency, projectedTransactionSize);
15    │ totalFrequency ← 0;
16    │ totalProjectedData ← 0;
17    │ foreach (frequency, projectedData) ∈ values do
18    │   │ totalFrequency ← totalFrequency + frequency;
19    │   │ totalProjectedData ←
      │   │ totalProjectedData + projectedTransactionSize;
20    │ if totalFrequency ≥ α * |D| then
21    │   │ if totalProjectedData ≤ γ then
22    │   │   │ Emit(null, (X, +)) // X belongs to T_k^+ ;
23    │   │ else
24    │   │   │ Emit(null, (X, −)) // X belongs to T_k^− ;
```

4.4   Local Mining of Frequent Itemsets

As described in the previous sections, the two-phase mining strategy guarantees the efficiency of MapFIM+. Indeed, when each projected-database stemming from a prefix is sufficiently small to be handled by a single node in the cluster (i.e., $\|D'_X\| \leq \gamma$), MapFIM+ switches to the local mining phase. After presenting the local mining method, we will show how to configure γ parameter to ensure that this method has sufficient memory space for D'_X processing.

Method. In the local mining phase, the frequent itemset enumeration is completed by using a traditional efficient algorithm (for instance, Eclat [20] or LCM [19]) that fits the memory constraints required by single nodes. More precisely, this algorithm has to enumerate all the itemsets corresponding to a given prefix X in a linear memory space with respect to the size of D'_X. Level-wise algorithms will therefore not be adapted since it is difficult to limit themselves to a given prefix and the amount of memory required is very variable. Similarly, approaches based on FP-trees do not guarantee a bounded amount of memory for tree storage. However vertical database layout based approaches such as Eclat or LCM fit well the requirement of bounded memory usage.

Algorithm 4 details this step, which is still MapReduce driven. Local memory-fitted projected-databases are dispatched to each node (as Reducers) that allow to run the selected local FIM algorithm. Due to the difference in size among projected databases, the local mining could lead to a load imbalance among reducers. In [15], the authors of BigFIM algorithm have experimented different strategies to assign the prefixes and it is shown that a random method can achieve a good workload balancing.

Algorithm 4. MapFIM: Local Mining

---

1 **Function** LocalMap(*String key, String value*):
2   // *key*: input name;
3   // *value*: a subset of transactions in $\mathcal{D}$;
4   **while** *there are unsolved locally tractable itemsets* **do**
5         $\mathcal{T}^+_{Block} \leftarrow$ A block of $\mathcal{T}^+$ in HDFS;
6   **foreach** *itemset* $X \in \mathcal{T}^+_{Block}$ **do**
7             $i_{max} \leftarrow max_{<_{\mathcal{I}}}(X)$ // The last item in X;
8   **foreach** *transaction* $t \in value$ *that contains* X **do**
9                 $t' \leftarrow \{j \in t : i_{max} <_{\mathcal{I}} j\}$;
10   **if** $t' \neq \emptyset$ **then**
11   Emit(X, t');

12 **Function** LocalReduce(*String key, Iterator values*):
13   // *key*: an itemset X;
14   // *values*: the projected database $\mathcal{D}_X$ of X;
15   Create an empty file f_{in} in local disk;
16   Save *values* into f_{in};
17   Run a local FIM program with input=f_{in}, output=f_{out}, support=$\alpha * \frac{|\mathcal{D}|}{|values|}$;
18   **foreach** *frequent itemset* $X' \in f_{out}$ **do**
19         $X'' = X \cup X'$;
20   Emit(*null,X''*) // X'' *belongs to* $\mathcal{L}$;

In the Map phase (lines 1–11 in Algorithm 4), we consider frequent itemsets $X \in \mathcal{T}^+$ as prefixes and construct their projected databases. For each $X \in \mathcal{T}^+$, i_{max} denotes the maximal item in X w.r.t $<_{\mathcal{I}}$. The projected-database $\mathcal{D}'_X$ is built by: (1) pruning every transaction $t \in \mathcal{D}'$ that does not contain X (2) pruning every item $j \leq_{\mathcal{I}} i_{max}$ since these items cannot expand X due to the prefix-based join. As shown in Algorithm 4, each Mapper reads a block of data, then for each $X \in \mathcal{T}^+$, it emits every transaction t' that contains X after pruning unnecessary items (line 11).

In the Reduce phase (lines 12–20 in Algorithm 4), a local FIM algorithm is independently called to enumerate all the frequent itemsets for each projected-database. More precisely, in the Reduce phase, each key is a frequent itemset $X \in T^+$ and each list of $values$ contains all transactions of the projected-database of X. They are saved to a local file so that the local FIM algorithm can work on it. For each itemset X' being frequent in the projected-database, the itemset $X'' = X \cup X'$ is frequent in $\mathcal{D}$. Notice that in the case where T^+ is too large to fit in memory of Mappers, we partition this set into several parts and repeat a local mining (via a MapReduce phase) for each part until that all itemsets in T^+ are processed.

Automatic Parameterization of MapFIM+. In our algorithm, a good value of γ is important for getting high performance. The higher the value of γ is, the better performance we get in general but more memory is required. Unfortunately the configuration of this threshold is complex for a user. Indeed, the appropriate choice of this parameter requires a good understanding of the approach in order to anticipate the amount of memory to consume at the level of this local mining phase. Without an automatic configuration system, a novice user could make the extraction impossible (memory saturation or extraction time too long). Even a more seasoned user might not get the best of the approach.

The automatic procedure for parameterizing the minimum size of the projected dataset γ requires an offline calibration phase which is carried out only once before extractions. The first step is to select a local frequent itemset mining algorithm whose memory is proportional to the size of the projected database as explained above. For instance, in our experiments, we use a local program based on Eclat/LCM algorithm [19, 20] which requires a maximum of $f(projectedData)$ of memory, where $f()$ is a linear function. Then, the maximum memory needed by the program is $K \times projectedData$ where K is the amount of memory required per item of $projectedData$ in the worst case during the extraction. The second step is to calibrate this algorithm by applying it to various datasets and then, figuring out the value of K. Once the constant K is known, it is easy to configure the parameter γ (whatever the dataset and the minimum frequency threshold) by applying a proportionality law, taking care to keep enough memory for performing the reducer. More precisely, we propose to set the threshold γ in MapFIM+ as follows:

$$\gamma = \frac{M_{Reduce} - M_{reduce\_task}}{K} \qquad (1)$$

where M_{Reduce} is the limit of memory of a Reducer and M_{reduce_task} is the memory required for a reduce task without running the local mining program.

4.5   Completeness and Extensibility

Thanks to the complementarity of global and local mining phases, this section demonstrates that MapFIM+ is not only correct and complete, but also transaction-extensible.

Let us first recall that MapFIM+ is composed of two phases: a global phase computes $\mathcal{G}$, the set of frequent itemsets that are either not *locally tractable* or that are *minimally locally tractable*, then a local mining phase computes the frequent itemsets derived from frequent *minimally locally tractable* itemsets. We show that the global phase is complete w.r.t. $\mathcal{G}$ and that the whole algorithm is complete w.r.t. the frequent itemsets, under the conditions of the *prefix-completeness* of the local algorithm.

Proposition 1. *MapFIM+ is **correct**, i.e., all itemsets returned by the algorithm are frequent and **complete**, i.e., all frequent itemsets are returned by the algorithm.*

Proof. The algorithm counts the support of each itemset and returns only frequent itemsets, therefore it is correct. The proof of the completeness is decomposed in two steps: proof of the completeness of the global phase w.r.t. $\mathcal{G}$ and the proof of the completeness of MapFIM+ w.r.t. $\mathcal{L}$. In the following, let $X = (i_1, \dots, i_k)$ be an itemset with $i_1 < i_2 \dots < i_k$, then X_j denotes the prefix of X, with length j, i.e., $X_j = (i_1, \dots, i_j)$.

Proof of the completeness of the global phase w.r.t. $\mathcal{G}$. We first prove by recurrence on k that the algorithm is complete w.r.t. $\mathcal{G}$, i.e. the set of frequent *minimally locally tractable* itemsets and of frequent non *locally tractable* itemsets. We recall that for $k > 1$, we have $\mathcal{C}_k = \mathcal{T}_{k-1}^- \bowtie \mathcal{G}_{k-1}$. Therefore, to prove completeness, we have to prove that $\mathcal{G}_k \subseteq \mathcal{C}_k$.

This is true for $k = 1$, since during data preparation, the support of all items are counted and only non frequent items are discarded. Thus, it allows to compute $\mathcal{G}_1$. Now, let us suppose that the algorithm is complete w.r.t. $\mathcal{G}_{k-1}$ and let us show that the algorithm is complete w.r.t. $\mathcal{G}_k$.

Let $X = (i_1, \dots, i_k)$, with $i_1 < i_2 \dots < i_k$ be an element of $\mathcal{G}_k$ and let us show that $X \in \mathcal{C}_k$. X is equal to the join of two $k-1$ itemsets, $(i_1, \dots, i_{k-1})$ and $(i_1, \dots, i_{k-2}, i_k)$, i.e. $X = (i_1, \dots, i_{k-1}) \bowtie (i_1, \dots, i_{k-2}, i_k)$. We have two cases:

- $X \in \mathcal{T}_k^-$: X is not *locally tractable*, i.e. $supp(X, \mathcal{D}) \geq \alpha$ and $\|\mathcal{D}_X\| > \gamma$.
    - $(i_1, \dots, i_{k-1}) \in \mathcal{T}_{k-1}^-$, since it is frequent and it is not *locally tractable* (as a prefix of an itemset that is not *locally tractable*)
    - $(i_1, \dots, i_{k-2}, i_k)$ is frequent but it is not a prefix of X. It is either not *locally tractable* or *locally tractable*. If it is *locally tractable*, it is *minimally locally tractable*, since its prefix is not *locally tractable*. Therefore it belongs either to $\mathcal{T}_{k-1}^-$ or to $\mathcal{T}_{k-1}^+$.

    Therefore $(i_1, \dots, i_{k-1}) \in \mathcal{T}_{k-1}^-$ and $(i_1, \dots, i_{k-2}, i_k) \in \mathcal{T}_{k-1}^- \cup \mathcal{T}_{k-1}^+ = \mathcal{G}_{k-1}$ and X belongs to $\mathcal{T}_{k-1}^- \bowtie \mathcal{G}_{k-1}$

- $X \in \mathcal{T}_k^+$: X is *minimally locally tractable*, i.e. $supp(X, \mathcal{D}) \geq \alpha$, $\|\mathcal{D}_X\| \leq \gamma$, and all its prefixes X_j satisfy $\|\mathcal{D}_{X_j}\| > \gamma$
    - $(i_1, \dots, i_{k-1})$ and $(i_1, \dots, i_{k-2})$ are frequent but not *locally tractable* (otherwise X would not be *minimally locally tractable*), i.e. $(i_1, \dots, i_{k-1}) \in \mathcal{T}_{k-1}^-$ and $(i_1, \dots, i_{k-2}) \in \mathcal{T}_{k-2}^-$
    - $(i_1, \dots, i_{k-2}, i_k)$ is either in $\mathcal{T}_{k-1}^-$ or in $\mathcal{T}_{k-1}^+$ (since $(i_1, \dots, i_{k-2}) \in \mathcal{T}_{k-1}^-$), i.e. $(i_1, \dots, i_{k-2}, i_k) \in \mathcal{G}_{k-1}$

As a consequence, X belongs to $\mathcal{T}_{k-1}^- \bowtie \mathcal{G}_{k-1}$.

Proof of the completeness of MapFIM+ w.r.t. $\mathcal{L}$.

Now we can prove the completeness of the algorithm w.r.t. the set $\mathcal{L}$ of frequent itemsets. Let $X = (i_1, \ldots, i_k)$, with $i_1 < i_2 \ldots < i_k$, be a frequent itemset. We have two cases:

- $X \in \mathcal{T}^-$, therefore $X \in \mathcal{G}$ and we have already shown the completeness of the algorithm w.r.t $\mathcal{G}$.
- $X \notin \mathcal{T}^-$: $supp(X, \mathcal{D}) \geq \alpha$ and $\|\mathcal{D}_X\| \leq \gamma$. Let j be the smallest index such that $X_j \in \mathcal{T}^+$. X_j belongs to $\mathcal{G}$ and therefore it has been generated in the global phase. If $j = k$ then $X_j = X$ has been generated at the global phase, otherwise *under the condition that the local mining algorithm is prefix complete*, X is generated in the local phase. More precisely, the frequent itemsets starting by X_j will be mined in the local mining step, from the conditional database with respect to X_j. It is built by considering all transactions in $\mathcal{D}$ containing X_j and removing from these transactions all items i with $i \leq i_j$. Since X is ordered, if X is frequent in $\mathcal{D}$ then $\{i_{j+1}, \ldots, i_k\}$ is frequent in the conditional database w.r.t X_j and will be found during the local mining phase.

The main challenge faced by MapFIM+ is to deal with a very large number of transactions. This is possible because the preparation and the scanning of this transactional database is distributed on several mappers and the set of generated candidates that is potentially huge is stored on the distributed file system. Therefore, in addition to being complete, MapFIM+ is transaction-extensible as introduced by Definition 2:

Proposition 2 (Transaction-extensible). *Assuming the distributed file system has an infinite storage capacity, MapFIM+ is transaction-extensible when the set of items I holds in memory and the local frequent itemset mining method takes space $O(\gamma)$.*

Proof: The first step of data preparation is not a problem as it is similar to MapReduce *WordCount* problem. The second step is also transaction-extensible because the set of frequent items holds in memory as we make the assumption that the set of all items holds in memory. Global mining phase does not raise any problem because all candidates are stored on the distributed file system (which has an infinite storage capacity) and can be partitioned into independent blocks of candidates (see line 5 of Algorithm 3). For local mining phase, the minimally locally tractable itemsets are also considered block by block (see line 5 of Algorithm 4). In the reduce step, the mining algorithm for a prefix takes a memory space proportional to the size of its projected database so there is at least one γ such that each projected database holds in memory.

5   Experiments

The experimental evaluation mainly focuses on performance and transaction-extensibility of the proposed Memory Aware Parallelized Frequent Itemset method. The research questions are as follows:

- **Q1** How MapFIM+ compare to MapFIM, MapFIM+ being an improved version of our previous MapFIM method?
- **Q2** How transaction-extensible is MapFIM+, i.e. does it manage to mine all frequent itemsets whatever the number of transactions in the dataset and the minimum support threshold?
- **Q3** How MapFIM+ compare to the best approaches for itemset mining using Hadoop MapReduce framework, in particular BigFIM and PFP?

Question Q1 is addressed in Sect. 5.2, whereas questions Q2 and Q3 are both addressed in Sect. 5.3. All the experiments were performed on a cluster of 16 virtual machines, where each virtual machine possesses 4 vCPUs, 8 GB RAM, and 300 GB HDD space. Each map/reduce task is allowed to use up to 7 GB of RAM. MapFIM+, MapFIM and PFP are experimented on top of Hadoop 2.7.3 while BigFIM is tested on Hadoop 1.2.1.[2]

5.1   Experimental Setup

Data Sets. In our experiments, we have chosen WebDocs dataset [11], one of the largest commonly used datasets in Frequent Itemset Mining. It is derived from real-world data and has a size of 1.48 GB. The copy of the dataset used in our experiments is obtained from the Frequent Itemset Mining Implementations Repository at http://fimi.ua.ac.be/data/.

We have also generated various synthetic datasets by using the generator from the IBM Almaden Quest research group. Their program can no longer be downloaded and we have used another repository available at https://github.com/zakimjz/IBMGenerator. The command used to generate our synthetic datasets is: ./gen lit -ntrans 50000 -tlen L -nitems 100 -npats 1000 -patlen 4 -ascii where L is the average length of transactions. We varied parameter L from 20 to 100 to generate 5 different datasets. The characteristics of the datasets are given in Table 4.

Setting of MapFIM+. For setting MapFIM+, we first apply the calibrating protocol described in Sect. 4.4 on the selected local FIM implementation based on Eclat/LCM algorithm [19,20]. With 10 datasets coming from the FIMI repository at http://fimi.ua.ac.be/data/, we run the program with a minimum support threshold equal to 0% to report the maximum memory used during one hour by the program. Then we compute $K = \frac{max\_memory}{projectedData}$ (in KB per item) and report the result in Table 5 (e.g., the value of K for dataset WebDocs is 0.018). From

---

[2] In our configuration, there is no real difference of performance between Hadoop 1.2.1 and Hadoop 2.7.3.

Table 4. Characteristic of the two used datasets

Dataset	Avg length	# items	# transactions	FileSize (GB)
WebDocs	177	5,267,656	1,692,082	1.5
T20.I100K.D50M	20	100,000	50,000,000	6.0
T40.I100K.D50M	40	100,000	50,000,000	11.9
T60.I100K.D50M	60	100,000	50,000,000	17.8
T80.I100K.D50M	80	100,000	50,000,000	23.7
T100.I100K.D50M	100	100,000	50,000,000	29.6

experiments, the value of K varies from 0.017 to 0.043, with an average value K of 0.023 and a standard deviation of 0.00785. For estimating γ, we fixed the value of K equals to 0.05. Finally, as the available memory space of the reducer is M_{Reduce} KB and the reduce task requires around M_{reduce_task} KB, we configure MapFIM+ by setting MapFIM+ such that: $\gamma = (M_{Reduce} - M_{reduce\_task})/0.05$.

Table 5. Parameter K setting using Borgelt's implementation of Eclat/LCM

Dataset	$projectedData$	max_memory (KB)	K (KB/item)
accidents	11,500,870	228,400	0.020
connect	2,904,951	58,160	0.020
kosarak	8,019,015	193,644	0.024
pumsb	3,629,404	63,332	0.017
retail	908,576	25,588	0.028
T40I10D100K	3,960,507	71,988	0.018
T10I4D100K	1,010,228	25,916	0.026
chess	118,252	5,100	0.043
pumsb_star	2,475,947	47,424	0.019
webdocs	299,887,139	5,422,024	0.018

5.2 Difference Between MapFIM+ and MapFIM

In this subsection, we focus on the first question **Q1**: How MapFIM+ compare to MapFIM? MapFIM+ is implemented similarly to MapFIM algorithm, both are in Java 8 for Hadoop 2. Moreover, for the local mining phase of both MapFIM+ and MapFIM, Eclat/LCM program implemented in C++ by Borgelt at http://www.borgelt.net/eclat.html was used to mine local projected database.

Experimental Setup. The main difference between MapFIM+ and MapFIM is how each approach estimates the projected datasets, based on the memory allocation for the local mining phase. Therefore, we experiment the two algorithms

Table 6. Estimated value of β parameter in MapFIM

Memory available for local mining program (MB)	Estimated β in MapFIM
1024	24%
2048	47%
3072	71%
4096	94%

using different values of the bound of memory available for the local mining program. Both programs are tested using WebDocs dataset with a value of support equals to 5%. We varied the bound of memory for local mining program from 1024 MB to 4096 MB.

For MapFIM+ algorithm, we recall that γ parameter is defined using the value of K equal to 0.05. For MapFIM algorithm, memory control is based on β parameter which is the number of transactions that can be processed locally. This β parameter is more difficult to set because the size of transactions in a projected database varies. A suboptimal solution is to bound the size of the projected database by using the maximum transaction length l: $\|\mathcal{D}'_X\| \leq \beta \times l$. By injecting this approximation into Eq. 1, we obtain for the threshold: $\beta = \gamma/l$. Table 6 presents the estimated value of β parameter in MapFIM.

For example, when memory available for the local program is bounded by 1024 MB, MapFIM estimates that local program can only handle projected datasets with at most 24% of the total transactions. With 4096 MB of available memory or more, the performances of MapFIM+ and MapFIM are identical because every projected datasets generated from frequent items can be mined locally. As a consequence, the number of prefix-projected datasets is the same in the two approaches. However, with limited memory, the difference between MapFIM+ and MapFIM is clear.

Experimental Results. Figure 2 shows execution time of the two programs (in seconds). As expected, with 4096 MB of available memory for allocation, there is no difference between the two programs in term of performance. However, for lower memory available for allocation, MapFIM+ performs better than MapFIM. Figure 3 shows the number of projected datasets generated by MapFIM+ and MapFIM for different cases of available memory of local processing nodes. By calculating the exact size of each projected dataset, MapFIM+ achieves a good performance compared to MapFIM by generating a smaller number of projected datasets. Not difficult to see, the gap between MapFIM+ and MapFIM shall be more important in the case of huge candidate sets due to the parallel candidate sets generation in MapFIM+, which is performed by a single processing node in MapFIM.

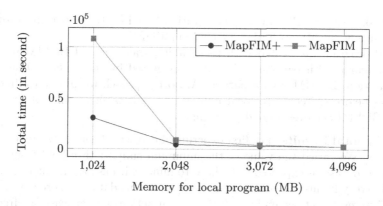

Fig. 2. Execution time of MapFIM+ and MapFIM using WebDocs dataset with support = 5%

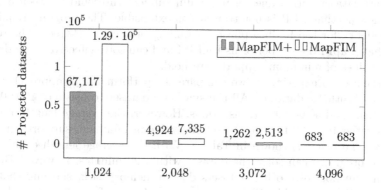

Fig. 3. Number of projected datasets using different bounds of local memory (MB)

5.3 Comparison to Other Existing FIM Approaches

In this subsection, we focus on the second and third questions:

- **Q2** How transaction-extensible is MapFIM+?
- **Q3** How MapFIM+ compare to the best existing approaches for itemset mining using Hadoop MapReduce framework?

Experimental Setup. Beside MapFIM+, we believe that Parallel FP-Growth (PFP) [8] and BigFIM algorithms [15] are the best approaches for itemset mining using Hadoop MapReduce framework. Thus, we decide to compare the performance of MapFIM+ to BigFIM and PFP algorithms. Moreover, we check that MapFIM+ is transaction-extensible, meaning that it can mine all frequent itemsets whatever the number of transactions in the data set and the minimum support threshold, which is not the case for PFP and BigFIM algorithms. In our experiments, we use PFP implementation available in the library Apache

Mahout 0.8 [12] and BigFIM implementation based on Hadoop 1 and provided by the authors at https://gitlab.com/adrem/BigFIM-sa.

PFP program was tested with its default parameter and BigFIM program was configured with parameter $k = 3$ as suggested by the authors. Using this configuration, BigFIM uses a parallel Apriori approach to mine all 3-frequent itemsets before switching to local mining phase. It is shown in [15] that with $k = 3$, BigFIM achieves a good performance.

Experimental Results. The first experiment consists of testing the real dataset WebDocs. We varied the value of the minimum support threshold from 5% to 15%. This dataset is expected to be hard to mine as it has long frequent itemsets as well asvery frequent itemsets. For example, in this dataset, there exists a frequent 7-itemset that occurs in 20% of the transactions and at least one frequent 3-itemset that appears in more than 60% of transactions.

The results are shown in Table 7. It is surprising that PFP can not solve WebDocs dataset with values of minimum support threshold between 5% and 15%, showing that PFP is not transaction-extensible. The program requires a huge memory for the Reduce phase for mining projected FP-trees. The results show that MapFIM+ outperforms BigFIM and can solve effectively the dataset for low values of minimum support threshold.

In the second experiment, we compare the performance of approaches with generated synthetic datasets. All the synthetic datasets consists of 100.000 different items and 50.000.000 transactions. However, the average length of transactions varies from 20 to 100. We set the value of minimum support threshold to 0.2%. The result is shown in Table 8. With the average length equal to 20, all three programs can solve the dataset within 20 minutes. However, BigFIM program can not solve other datasets due to memory lack, showing that it is not transaction-extensible. PFP program can solve two further datasets but not the last ones where the average length of transactions is equal to 80 and 100. It is clear that MapFIM+ has a better execution time and is able to solve harder datasets with longer transactions. These results confirm that MapFIM+ is transaction-extensible.

From the papers presenting the different approaches and the implementations of the programs, in our opinion, there are three main reasons that explain why MapFIM+ achieves better performances:

- Our approach generates balanced prefix-projected datasets in an efficient manner while guaranteeing that projected datasets can always be mined locally. This makes local mining of projected datasets more efficient compared to existing approaches.
- Different from BigFIM and PFP, MapFIM+ does not implement the local mining program but is able to use any efficient FIM implementations such as Eclat, so the performance of MapFIM+ could be further improved while applying additional optimizations to local mining process.
- MapFIM+ is transaction-extensible while guaranteeing good balancing properties, among processing nodes, during all computation steps.

Table 7. Execution time (in seconds) using WebDocs dataset

Support	MapFIM+	BigFIM	PFP
15	392	4136	Out of Memory
14	401	5583	Out of Memory
13	446	8207	Out of Memory
12	465	12319	Out of Memory
11	514	19748	Out of Memory
10	615	32267	Out of Memory
9	703	Out of Memory	Out of Memory
8	894	Out of Memory	Out of Memory
7	1208	Out of Memory	Out of Memory
6	1798	Out of Memory	Out of Memory
5	2684	Out of Memory	Out of Memory

Table 8. Execution time (in seconds) using synthetic datasets

Dataset	MapFIM+	BigFIM	PFP
T20.I100K.D50M	788	854	1009
T40.I100K.D50M	3178	Out of Memory	10838
T60.I100K.D50M	7623	Out of Memory	54031
T80.I100K.D50M	15092	Out of Memory	Out of Memory
T100.I100K.D50M	24749	Out of Memory	Out of Memory

6 Conclusion and Future Work

In this paper, we present MapFIM+ an improved version of our previous Map-
FIM algorithm [7], a MapReduce based two-phase approach to efficiently mine
frequent itemsets in very large datasets. In the first global mining phase, MapReduce is used to generate local memory-fitted prefix-projected databases from
the input dataset benefiting from the Apriori principle. Then, in a local mining phase, an optimized in-memory mining process is launched to enumerate in
parallel all frequent itemsets from each prefix-projected database. Compared to
other existing approaches, our algorithm implements a fine-grained method to
switch from global phase to the local mining phase. Moreover, we show that
our method is transaction-extensible, meaning that given a fixed set of items,
it can mine all frequent itemsets whatever the number of transactions and the
minimum support threshold. To the best of our knowledge, our algorithm is the
first one to guarantee this property.

Our experimental evaluations show that both MapFIM and MapFIM+ outperform the best existing MapReduce based frequent itemset mining approaches.

Moreover, MapFIM+ performs better than MapFIM in the case of huge candidate sets by reducing communication and disks I/O costs. We show how to define and set the unique γ parameter of our MapFIM+ algorithm depending only on the available memory for local mining using prefix-projected databases. This point is particularly important, since an optimal value of γ parameter guarantees a high performance level.

Future work will be devoted to make MapFIM+ scalable. This can be achieved by using similar approaches, as those based on randomized key redistributions and introduced in [3,4] for join processing. Indeed, the use of randomized key redistributions prevents the effects of data skew while guaranteeing perfect balancing properties during all the stages of join computation in large scale systems.

Acknowledgement. This work is partly supported by the GIRAFON project funded by *Centre-Val de Loire* region (France).

References

1. Aggarwal, C.C., Han, J.: Frequent Pattern Mining. Springer, Cham (2014). https://doi.org/10.1007/978-3-319-07821-2
2. Agrawal, R., Srikant, R., et al.: Fast algorithms for mining association rules. In: Proceedings of VLDB 1994, vol. 1215, pp. 487–499 (1994)
3. Al Hajj Hassan, M., Bamha, M.: Towards scalability and data skew handling in groupby-joins using MapReduce model. In: Proceedings of ICCS 2015, pp. 70–79 (2015)
4. Al Hajj Hassan, M., Bamha, M., Loulergue, F.: Handling data-skew effects in join operations using MapReduce. In: Proceedings of ICCS 2014, pp. 145–158. IEEE (2014)
5. Beedkar, K., Berberich, K., Gemulla, R., Miliaraki, I.: Closing the gap: sequence mining at scale. ACM Trans. Database Syst. **40**(2), 8:1–8:44 (2015)
6. Dean, J., Ghemawat, S.: MapReduce: simplified data processing on large clusters. Commun. ACM **51**(1), 107–113 (2008)
7. Duong, K.-C., Bamha, M., Giacometti, A., Li, D., Soulet, A., Vrain, C.: Map-FIM: memory aware parallelized frequent itemset mining in very large datasets. In: Benslimane, D., Damiani, E., Grosky, W.I., Hameurlain, A., Sheth, A., Wagner, R.R. (eds.) DEXA 2017. LNCS, vol. 10438, pp. 478–495. Springer, Cham (2017). https://doi.org/10.1007/978-3-319-64468-4_36
8. Li, H., Wang, Y., Zhang, D., Zhang, M., Chang, E.Y: PFP: parallel FP-growth for query recommendation. In: Proceedings of RecSys 2008, pp. 107–114. ACM (2008)
9. Li, N., Zeng, L., He, Q., Shi, Z.: Parallel implementation of apriori algorithm based on MapReduce. In: Proceedings of SNDP 2012, pp. 236–241. IEEE (2012)
10. Lin, M.-Y., Lee, P.-Y., Hsueh, S.-C.: Apriori-based frequent itemset mining algorithms on MapReduce. In Proceedings of ICUIMC 2012, pp. 76:1–76:8 (2012)
11. Lucchese, C., Orlando, S., Perego, R., Silvestri, F.: WebDocs: a real-life huge transactional dataset. In: FIMI, vol. 126 (2004)
12. Apache Mahout: Scalable machine learning and data mining (2012)
13. Makanju, A., Farzanyar, Z., An, A., Cercone, N., Hu Z.Z., Hu, Y.: Deep parallelization of parallel FP-growth using parent-child MapReduce. In: Proceedings of BigData 2016, pp. 1422–1431. IEEE (2016)

14. Miliaraki, I., Berberich, K., Gemulla, R., Zoupanos, S.: Mind the gap: large-scale frequent sequence mining. In: Proceedings of SIGMOD 2013, pp. 797–808. ACM (2013)
15. Moens, S., Aksehirli, E., Goethals, B.: Frequent itemset mining for big data. In Proceedings of BigData 2013, pp. 111–118. IEEE (2013)
16. Salah, S., Akbarinia, R., Masseglia, F.: Data partitioning for fast mining of frequent itemsets in massively distributed environments. In: Chen, Q., Hameurlain, A., Toumani, F., Wagner, R., Decker, H. (eds.) DEXA 2015. LNCS, vol. 9261, pp. 303–318. Springer, Cham (2015). https://doi.org/10.1007/978-3-319-22849-5_21
17. Salah, S., Akbarinia, R., Masseglia, F.: Optimizing the data-process relationship for fast mining of frequent itemsets in mapreduce. In: Perner, P. (ed.) MLDM 2015. LNCS (LNAI), vol. 9166, pp. 217–231. Springer, Cham (2015). https://doi.org/10.1007/978-3-319-21024-7_15
18. Savasere, A., Omiecinski, E., Navathe, S.B.: An efficient algorithm for mining association rules in large databases. In: Proceedings of VLDB 1995, pp. 432–444. Morgan Kaufmann Publishers Inc., San Francisco, CA, USA (1995)
19. Uno, T., Asai, T., Uchida, Y., Arimura, H.: LCM: an efficient algorithm for enumerating frequent closed item sets. In: FIMI, vol. 90. Citeseer (2003)
20. Zaki, M.J., Parthasarathy, S., Ogihara, M., Li, W., et al.: New algorithms for fast discovery of association rules. In: KDD, vol. 97, pp. 283–286 (1997)
21. Zhou, L., Zhong, Z., Chang, J., Li, J., Huang, J.Z., Feng, S.: Balanced parallel FP-growth with MapReduce. In: Proceedings of YC-ICT 2010, pp. 243–246 (2010)

Author Index

Printed in the United States
By Bookmasters